Fabricating Printed Circuit Boards

Fabricating Printed Circuit Boards

Jon Varteresian

An imprint of Elsevier Science

Amsterdam   London   New York   Oxford   Paris   Tokyo
Boston   San Diego   San Francisco   Singapore   Sydney

Newnes is an imprint of Elsevier Science.

∞ Recognizing the importance of preserving what has been written, Elsevier Science prints its books on acid-free paper whenever possible.

Library of Congress Cataloging-in-Publication Data

Varteresian, Jon.
   Fabricating printed circuit boards / Jon Varteresian.
      p. cm.
   Includes index.
   ISBN: 1-878707-50-7 (pbk. : alk. paper)
   1. Printed circuits—Design and construction. I. Title.

TK7868.P7 V27 2002
621.3815'31—dc21   2002141586

British Library Cataloguing-in-Publication Data
A catalogue record for this book is available from the British Library.

The publisher offers special discounts on bulk orders of this book.
For information, please contact:

Manager of Special Sales
Elsevier Science
225 Wildwood Avenue
Woburn, MA 01801-2041
Tel: 781-904-2500
Fax: 781-904-2620

For information on all Newnes publications available, contact our World Wide Web home page at: http://www.newnespress.com

10 9 8 7 6 5 4 3 2 1

Printed in the United States of America

Dedication

For Katie and Emma, my brightest stars.

Contents

Introduction

How difficult is it to design and fabricate printed circuit boards?

When you look at a finished printed circuit board—with an often-complex circuit pattern and a mixture of through-hole and surface mount components—you may think creating your own boards would be a difficult, time-consuming task that would require specialized tools and expertise. However, fabricating your own printed circuit boards can be broken down into the following relatively simple steps:

1. Generation of the schematic.

2. Placement and routing of the circuit board.

3. Generation of artwork.

4. Exposing and developing the resist layer.

5. Etching the printed circuit board.

6. Tin plating of the printed circuit board.

7. Drilling and shaping, including *vias* (or barrels).

Depending upon your particular situation, you may not need to complete all of the steps listed above. The purpose of this book is to explain each of the steps so you can create your own professional-quality printed circuit boards.

The method of producing printed circuit boards described in this book is unique. It is a result of years of trial and error and lots of ruined circuit boards! What makes this process unique is that you don't need photographic equipment, lots of strange and dangerous chemicals, or expensive traditional artworks. All artworks used in this process are printed on plain white paper—no transparencies or commercially produced photoplots are needed! You can generate these artworks directly from your own laser printer, ink jet printer, or copier. This process uses pre-sensitized positive circuit boards so your artworks should be a positive of your patterns (black where you want copper, blank where you don't) as compared to a negative that is the inversion of your patterns. This means you can now copy artworks from a printed source (such as a magazine, book, or data sheet) using a photocopier and use them directly, without any modifications, to produce quality circuit boards.

Most of the problems historically encountered and conquered during the fabrication of printed circuit boards have dealt with the generation of the artworks themselves. Your first thought for generating artworks might be to have them commercially photoplotted. A *photoplot* is nothing more than a high quality transparency, or viewgraph. However these can cost up to $20 to $40 per photoplot and be a significant item on project budgets.

Some people have tried to get around this problem by printing their artworks directly on transparencies, thus effectively simulating photoplots. However, those who have tried this soon realize that a transparency can stretch and slip as it proceeds through many printers and toner does not bond well to transparencies in small or thin areas. If you have a circuit board that is longer than 3 inches or so, your final product will not live up to your expectations. Multiple layer registration can also be difficult with artworks that are stretched.

You may have tried some of the new iron-on toner transfers that are now available. Our success rate with these products has been very low. The whole concept depends on perfect iron temperature and pressure. While a good idea in theory, we have found that sections of the artworks don't transfer due to iron pressure being too low while other sections smudged or smeared because of iron pressure being too high.

The solution we have developed is to print artwork directly on paper (less than 20-weight). This works great as far as layer registration goes since paper slips and stretches very little as it proceeds through a printer, and the exposure process still works because the paper is only slightly opaque. (Hold this page up to a bright light or window and you will notice that you can see through it.) The only drawback to using paper for artworks is that you need to expose the circuit board resist layer a little longer than usual. Don't try paper with a weight less than 16-weight (or something like vellum) because it too can slip and slide its way through a printer.

Why did we choose positive artworks instead of negative? Positive artworks look just like the patterns you are trying to create—black where you want copper, blank where you don't—as compared to a negative which is the inversion of the pattern you want. Negatives usually involve many black areas, and that's something printers usually have trouble producing. If you laser print a page with a large black rectangle on it, you will notice that the black tends to lighten as you approach the center of the square. This means touch-up work. The solution is to use positives since positives usually do not have many large black areas.

Safety

We can't stress safety enough! While the techniques described in this book are not inherently dangerous, you must wear appropriate protective clothing when required or exercise care and good judgment when handling chemicals or using required tools. In particular, ALWAYS wear gloves and eye protection whenever you work with chemicals and ALWAYS wear eye protection when drilling or machining a board.

The chemicals used in the processes described in this book are relatively safe; however, always follow the procedures below when using any chemical:

 ✎ Always store your chemicals in tightly sealed plastic or glass containers.

 ✎ Make sure each container is clearly labeled with the contents and the date it was stored.

 ✎ Don't use expired chemicals.

 ✎ Make sure your work area is properly ventilated and lighted.

 ✎ Make sure your work area is clean.

Your state and local community may have certain regulations concerning the storage and disposal of various chemicals. Always check with the appropriate local or state office regarding the disposal of any chemical. (In many towns, the fire chief is usually the hazardous waste coordinator.) Each chemical you purchase should come with a *material safety data sheet* (MSDS) describing its handling, disposal, and safety information. Read it and always follow its recommended procedures.

The Exposure Cone project contained in this book involves the use of electricity. If you are not experienced with electric wiring and safety procedures, enlist the help of a qualified, licensed electrician when wiring this project. This advice goes for all the procedures described in this book; if you have any doubts as to your ability to perform any step or use the required tools correctly, seek the assistance of qualified persons. If you're not certain how to do something, don't guess!

The resource list included in Chapter 7 lists all equipment and supplies you will need. Suggested sources as well as approximate prices are included whenever available. Keep in mind that the prices listed may change without notice. Contact the supplier for up-to-the-minute pricing and availability.

Fabricating Printed Circuit Boards

CHAPTER 1

Schematic Capture

You are probably very familiar with various schematic symbols and diagrams. A schematic is merely a collection of electronic symbols connected together with virtual "wires." The main reason you need a schematic when fabricating a printed circuit board is to provide input (a *netlist*) to your layout and routing tool. A netlist is a file, usually ASCII text, which defines the connections between the components in your design. Other uses for a schematic when fabricating boards include documentation, archiving, and the automatic generation of a bill of materials (BOM).

If your design is simple, you may not need to generate a schematic and it is possible to skip this step and proceed directly to layout and routing of the circuit board. However, you may still need to generate a netlist to feed to the layout tool.

The Schematic Symbol

For most circuits, a photograph or realistic drawing showing actual components and their interconnections would be far too complex to be of value in replicating the circuit. A *schematic symbol* is a simplified representation of a real-world component. A *schematic diagram* shows such representations of real-world components and a simplified "map" of how they are connected together. It would be silly to tape resistors and capacitors to a piece of paper, so instead we use drawings that represent those parts.

Schematic symbols come in all sizes and shapes. They can be as simple as stick figures or as complicated as a work of art, but they all have a few common features. These features are summarized below:

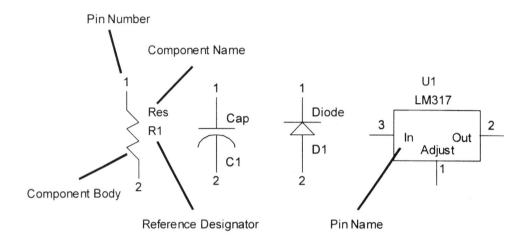

Figure 1-1: Examples of different schematic symbols

1) Each symbol contains a graphic, or drawing, that represents the part or its function. The graphic is usually referred to as a *component body*. In the examples shown in Figure 1-1 the graphics show a resistor, a capacitor, a diode, and an integrated circuit.

2) Each symbol needs to have a name that the schematic tool can refer to. This name is usually referred to as a *component name*. A component name for a symbol is similar in function to a file name for a file.

3) The component body also has lines coming off it called *pins* that represent the connections to the part itself.

4) Each pin has a unique pin number assigned to it. There should not be duplicate pin numbers on a single component. For example, the resistor shown should not have two pins numbered "2."

5) Each pin can have a name assigned to it. Each name should be unique for the symbol in question. The LM317 shown above has pin names associated with each pin.

6) Each symbol has a *reference designator* assigned to it in order to uniquely identify it from other symbols in the schematic. The schematic and layout tools typically use the reference designator to uniquely identify components in a design. The example reference designators shown are R1, C1, D1, and U1. Normally the reference designator is a letter (or letters) followed by a number. Duplicate reference designators should not be used in the same design.

Symbol Properties

Pin numbers, pin names, component names, etc. are referred to as *properties*. There are many other properties that can be attached to a symbol, such as part number, package name, manufacturer, description, etc. You can make up your own property names to suit your own individual goals. Whether you use these properties depends on whether your schematic capture tool can support them, how much detail you want on your schematic, and what your schematic and layout tools require. (For instance, most layout tools require a property that defines the component package type.)

After many years of generating schematics, I have come up with a list of properties that I feel properly describes a component. I consider these properties to be "musts." These properties and their descriptions are listed below. However, keep in mind that your schematic or layout tool may require other properties not listed below:

✏ *Part Number:* Describes the actual part number of the component (for example, "1N4148DICT"). If your schematic capture tool supports bill of materials (BOM) generation, this is what will be listed.

✏ *Component Name:* This is the name of the symbol from the schematic capture tool's point of view. Think of it as a file name for a symbol, such as "resistor," "resistor_surface_mount," or "diode."

✍ *Reference Designator:* Describes the reference name that the schematic and layout tools will use to identify the part. For example, you could place four resistors on a schematic. Each symbol may have the same component name (resistor_through_hole_.25w), but each will have a different reference designator (R1, R2, R3, and R4). Each symbol in a schematic must have a unique reference designator.

✍ *Package Name:* Describes the component's package type or footprint (for example, "DIP16" or "PLCC44").

✍ *Value:* Describes the component specifics, such as resistance or capacitance values, such as "240 ohms" or "40pf_10v." Not all symbols will need this property.

The following properties are not required but are nice to have:

✍ *Manufacturer:* Describes the manufacturer of the component for creating a bill of materials (BOM) for use when ordering components.

✍ *Description:* Describes the part to make it easier for others to understand (for example, "3:8 multiplexer").

Not every schematic tool will support all of the properties described above. You need to understand your schematic tool fully in order to decide which properties to use. Go through the examples and tutorials that come with your schematic and layout software to help you decide which properties are right for your application.

Schematic Generation

The first step in generating a schematic is to create symbols for all the components. The exact details of this process will vary with the schematic capture tool you choose. Most of the work prepared for this book was done with SuperCAD and SuperPCB by Mental Automation (see the resource list in Chapter 7). Once all the symbols are created, place or *instantiate* them on a blank schematic page or pages. It is a good idea not to fill the page up too tightly because you will not have room to add the wires, or connections, and any other parts you may decide you need later. Figure 1-2 shows a schematic with all the symbols instantiated but not connected. In this example, all the components fit on a single sheet. If you need more than one sheet, don't worry. Almost all of the schematic and layout tools can handle schematics drawn on multiple sheets. You can also increase the size of a single sheet if you prefer only one but this can make it difficult to handle, print and store.

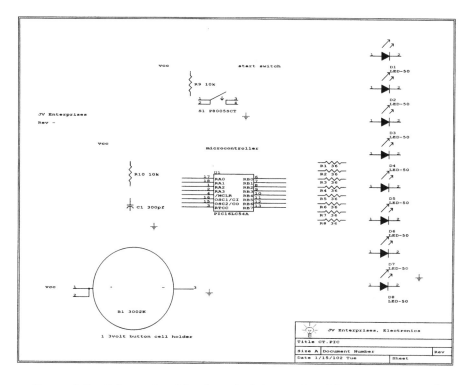

Figure 1-2: Schematic with all symbols instantiated but not connected

Once you have the symbols instantiated, it is time to start connecting them with "virtual wires." These "wires" are nothing more than lines drawn between the pins of the symbols. The schematic tool knows that these lines represent physical connections between the components. Figure 1-3 shows the same schematic with the connections completed. The exact details that you will follow will again vary with the schematic capture tool you choose. With all of them, however, it is as simple as activating the wire tool and clicking and dragging with the mouse.

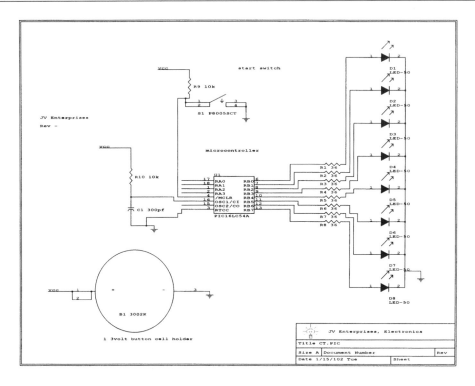

Figure 1-3: Schematic with all "wiring" connections made

For schematics with multiple sheets you will probably have to run wires, or *nets*, to other sheets. This is done with the net name property. Simply name the net on the first sheet, and then name any other nets on any other sheets that you want connected with the same name. The schematic tool will connect all nets with the same name. For example, a net named LED_ENABLE on one sheet will be electrically connected to all other nets in the design with the same name. Some schematic tools require you to place off page connectors on nets that leave and enter schematic sheets.

You should note that once a net has been named all other nets with that same name are connected together. Note that in Figure 1-3 there are three nets with the name VCC. All these nets are connected together even though there is no wire connecting them. Sometimes this is known as a *virtual connection*. This helps to keep any unnecessary clutter on a schematic sheet to a minimum.

Note that you do not have to name all of the nets in a schematic. The nets that you do not name will have a name assigned to them by the schematic tool. However, assigning names to all critical or important nets now will make it easier to refer to them when you are error checking the design, and later on when you are placing and routing the design.

You may have noticed that there are a few pins missing from the microcontroller, U1, shown in the previous figures. Specifically, those pins represent power and ground connections. Almost all integrated circuits contain power and ground connections that connect to a common supply rail such as VCC and GROUND. Why bother to add these pins to your symbols and clutter up the schematic? SuperCAD allows you to identify certain pins on a symbol as power and ground pins when you create the symbol. All power pins are connected to a power supply rail named VCC, and a ground rail named GROUND. SuperCAD automatically keeps track of these connections for you so you don't have to. In the schematic shown in Figure 1-3, the power connections to VCC and GROUND are implied for the microcontroller U1. However the connections are explicitly shown for the battery connector B1. Although not recommended, you can mix and match embedded and explicit power pins in a design.

This is a very nice feature, but beware! What happens if you have multiple power supply rails in your design, or you need separate analog and digital ground connections? Most schematic tools have some way in which to handle this but we would recommend adding the power and ground pins to your symbols and connecting them up explicitly. In fact, for most of the designs I've done, I've added all power and ground pins to the symbols so that their connections are obvious to all who look at the schematic. It also makes reviewing the design easier later on.

Once the schematic is complete you need to run a schematic design rule checker. Notice I said "need," not "should." Checking your design is a must. Design rule checkers vary greatly from schematic tool to schematic tool but they all try to do the same thing. They can look for unconnected nets, commonly referred to as *dangling* nets. They can look for shorted nets, unconnected pins, duplicated pins on a symbol, etc. Some tools like SuperCAD can also look for conflicts such as two or more output drivers on a single net, or no output drivers on a net. Read the documentation that came with your schematic tool.

Notice the comments placed on the schematic shown in Figure 1-3. These comments are not necessary but they will help you and anybody else who sees your schematic to understand it. Comments are free, so use them!

Generating a Netlist

Now that you have created a schematic, what do you do with it? Whether you know it or not, one of the main reasons for creating an electronic schematic was to help you with the placement and routing of the printed circuit board. (Not to mention giving your kids a lot of scrap paper to color on!) Once your schematic is complete and has passed all of the rules checking required by your schematic capture tool, you can generate a *netlist*. This netlist is usually an ASCII (text) document defining the connections shown in your schematic. This netlist is used by the layout tool help route your board.

The netlist for the schematic shown in Figure 1-3 is shown below. It was created with SuperCAD by Mental Automation.

Netlist for design: C:\JV_ENT\PCB\CT\CT

Time: 21:48

Date: 7/14/97 Mon

Processing time:6.4

Number of signals found=20

def0——d1-1a,r1-2z

def1——d2-1a,r2-2z

def2——d3-1a,r3-2z

def3——d4-1a,r4-2z

def4——r1-1z,u1-6u

def5——r2-1z,u1-7u

```
def6————r3-1z,u1-8u

def7————r4-1z,u1-9u

def8————r5-1z,u1-10u

def9————r5-2z,d5-1a

def10————r6-1z,u1-11u

def11————r6-2z,d6-1a

def12————r7-1z,u1-12u

def13————r7-2z,d7-1a

def14————r8-1z,u1-13u

def15————r8-2z,d8-1a

def16————u1-4i,r9-2z,s1-1u,s1-2u

def17————u1-16i,r10-2z,c1-1z

Vcc————u1-14,r9-1z,b1-1u,b1-2u,r10-1z

GROUND————u1-5,d1-2b,d8-2b,d2-2b,d3-2b,d4-2b,d5-2b,d6-2b,d7-
2b,u1-3i,b1-3u,&

    s1-3u,s1-4u,c1-2z
```

A typical netlist usually contains a header that lists the name of the design, time and date stamps and some various statistics such as total number of nets, processing time, etc. The main body of the netlist file contains all the connection information. For nets that do not have specific names, the schematic tool assigns names to them. That is what all the def# names in the above file are. After the net name, all of the pins connected to it are listed. For example net def5 consists of pin 1 of R2 and pin 7 of U1. This information will be fed to the layout tool to help insure that the circuit board traces match the schematic connections. If you were really ambitious, you could have created the above netlist by hand and saved the cost of the schematic tool but this really isn't a good idea.

It may look easy for simple designs, but for others with many pages and many more symbols the task will be immense and prone to errors.

You may notice other items in the netlist, such as the letters z,u,i and others. These letters define the direction of the pin in question. For example U1-3i tells us that Pin 3 of U1 is an input. Alternatively other pins may be defined as outputs (o), bi-directional (b), analog (a), or undefined (z). SuperCAD uses these directions to help check the design for errors. For example, there should not be a net in the design with more than one output connected to it. If you had the outputs of two logic gates connected together and one was a logic high while the other was a logic low, what would the state of that net be? After a while the state of that net would be broken! You also would not want a net without any outputs on it. The state of that net would be undefined (and probably useless). Each schematic tool handles these types of error checking differently. You should read the documentation that came with your software to determine just what kind of error checking, if any, your tools perform.

CHAPTER 2

Basic Circuit Board Placement and Routing Considerations

Once you have a schematic captured, error checked and netlisted, then what? The netlist you generated is fed into the circuit board layout tool and used to generate the artworks needed to fabricate the printed circuit board. The rest of this chapter will guide you through the steps involved.

Circuit Board Placement and Routing Basics

You can think of the circuit board layout process along similar lines as schematic capture. The main steps in capturing a schematic are:

1) Generating symbols

2) Placing the symbols

3) Connecting the symbols

Circuit board layout is very similar. The main steps are:

1) Generating component footprints

2) Placing footprints (placement)

3) Connecting the footprints (routing)

Typically the combination of placement and routing is referred to as *layout*.

Just what exactly is a component footprint? A *footprint* is a physical description of a component and is made up of padstacks, obstacles, and text. An example is shown in Figure 2-1. This footprint represents a sixteen-pin DIP package. Note that it is shown enlarged for clarity and that the drill holes are not shown.

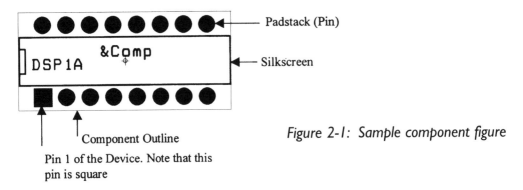

Padstack (Pin)

Silkscreen

Component Outline

Pin 1 of the Device. Note that this pin is square

Figure 2-1: Sample component figure

You can generate footprints that adhere to your company's standards, or create your own standard. Before we can get into the details, we need to define some terms. The terms are defined in their approximate order of appearance:

Obstacle: An outline or shape that represents an object on a circuit board that must be taken into account during placement or routing. The rectangle in the middle of the part shown in Figure 2-1 is a silkscreen obstacle. The outer rectangle is a component placement outline obstacle.

Layer: The definition of a layer is very important in the creation of footprints and padstacks. Padstacks are made up of obstacles that are placed on specific layers. In this way the user can control what connections are made on a specific layer of a circuit board. A circuit board can have many layers. A list of the more common ones is given below:

Top	Top or component layer
Bottom	Bottom or solder layer
Inner	All inner routing layers (inner1, inner2, etc.)
Plane	Power and ground planes
Solder mask top	Solder mask top
Solder mask bottom	Solder mask bottom
Solder paste top	Solder paste top
Solder paste bottom	Solder paste bottom
Silkscreen top	Silkscreen top
Silkscreen bottom	Silkscreen bottom

Assembly top	Assembly top
Assembly bottom	Assembly bottom
Drill Drawing	Drill Drawing
Drill	Drill holes and sizes
Fabrication Drawing	Fabrication drawing
Notes	Notes

On a single-layer circuit board there is only one layer of copper. This layer is commonly referred to as the *bottom or solder side*. Most layout tools can handle up to 16 copper layers. You can think of a multi-layer board as a stack of single layer boards all glued together. Drilling a hole through all the boards and inserting a metal *barrel*, or *via*, makes the connections between the layers. On these more complicated boards a padstack may have obstacles defined on one or more layers. There is no rule that a padstack must have obstacles defined on all layers in a circuit board. For example, a surface mount component will only have obstacles placed on the side of the circuit board on which it is placed, i.e., the top or bottom layers. Any inner layers will be blank, or *undefined*. This is because a surface mount component has no way to make connections to the inner layers of the circuit board. (There actually is a way to make connections to the inner layers, as you will see later on.) A through-hole component, such as a resistor, will have obstacles that are defined on all layers of the circuit board. This is because a hole had to be drilled through the board, and through each layer, to stick the pin of the component through.

Padstack: A padstack is a physical representation of a component's pin. In Figure 2-1 there are 15 round padstacks and one square. Typically the padstack for pin 1 of a footprint is different than the others. This makes it easier to align and orient the part when soldering it to the circuit board. A padstack is made up of one or more obstacles placed on specific layers. For example, a padstack could contain obstacles for the top layer of the circuit board, the inner layers, solder mask and paste mask. An example through-hole padstack layer stackup is shown in Figure 2-2. The exact dimensions of the obstacles on each layer will be discussed later.

Padstack or Layer Name	Pad Shape	Pad Width	Pad Height	X Offset	Y Offset
T1					
TOP	Round	62	62	0	0
BOTTOM	Round	62	62	0	0
PLANE	Round	70	70	0	0
INNER	Round	62	62	0	0
SMTOP	Round	67	67	0	0
SMBOT	Round	67	67	0	0
SPTOP	Undefined	0	0	0	0
SPBOT	Undefined	0	0	0	0
SSTOP	Undefined	0	0	0	0
SSBOT	Undefined	0	0	0	0
ASYTOP	Undefined	0	0	0	0
ASYBOT	Undefined	0	0	0	0
DRLDWG	Round	38	38	0	0
DRILL	Round	38	38	0	0
COMMENT LAYER	Round	38	38	0	0
SPARE2	Round	38	38	0	0
SPARE3	Round	38	38	0	0

Figure 2-2: Through-hole padstack definition, ORCAD

Via: In its simplest terms, a via is a connection between one layer of a circuit board and another. On a single-layer circuit board there isn't a need for vias, as all connections are done in copper on one layer. On a multi-layer circuit board, you have to be able to connect a trace on one layer to a trace on another layer. This is where the via comes in. Take a look at Figure 2-3. This figure shows an 8-pin surface mount device and two through-hole resistors. There are also two traces drawn on the top layer connecting U1-8 to R1-1 and U1-6 to R1-2. There is also a trace connected to U1-4 which goes elsewhere. Now, we want to connect a trace from U1-7 to R2-1, but there isn't a clear path to do this.

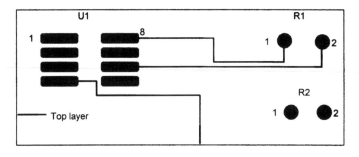

Figure 2-3: An example of a via

To make this connection we will use a via. The procedure is to draw a short trace from U1-7 on the top layer. Next, draw a trace from R2-1 on the bottom layer and line up the end with the short trace drawn on the top layer. It appears that we cross over a trace on the top layer but that is not the case since the two traces are on different layers. Once the two traces end in the same location, drill a hole, insert a metal barrel and solder both ends making the connection. See Figure 2-4 for details.

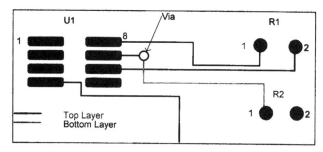

Figure 2-4: Continuation of the via example

Figures 2-5 and 2-6 show what the top and bottom layers each look like. Note that the through-hole component padstacks show up on all layers while the surface mount padstacks only show up on the top layer.

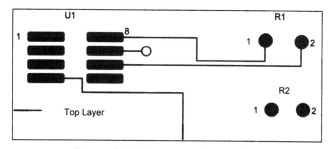

Figure 2-5: Via example top layer

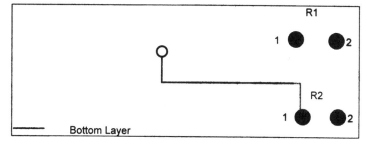

Figure 2-6: Via example bottom layer

Solder mask: In a homemade circuit board, the top layer (and bottom if double-sided) will consist of etched, bare copper. During the tin plating process described in this manual, all the copper will be plated with tin. This will help protect the copper from oxidation, and help increase its solderability. The only problems with tin plating are that it is expensive and leaves the circuit board susceptible to shorting. For a professionally fabricated circuit board we need something better. A fabrication house only tin plates the areas that you will actually solder to and areas that needs to be plated. Typically this includes all surface mount and through hole padstacks and vias. The solder mask layer is used to tell the fabrication house where to apply tin and where not to. The areas of the circuit board that do not get plated with tin are covered with a material known as solder mask. Solder mask is usually light green or blue, about 1 to 2 mils thick, and covers the bare copper to protect it from oxidizing as well as insulating it electrically. It is a relatively tough coating but will scratch. You place obstacles on the solder mask layer where you do not want solder mask. For example, an obstacle drawn on the solder mask layer will result in that area of the circuit board being tin plated, i.e., no solder mask.

Every time you make a padstack, you should always add a solder mask layer obstacle. Otherwise the pad will be covered with solder mask and you won't be able to solder the part to it. The solder mask obstacle should have the same size and shape of the top layer pad. Figure 2-7 shows an example of a section of a circuit board with surface mount components, through-hole components and some routing. Figure 2-8 shows the same area's solder mask layer. Note that all surface mount pads, through-hole pads, and vias have a corresponding obstacle on the solder mask layer and that the traces do not. When this circuit board is fabricated all pads and vias will be tinned while all the traces will be protected under a layer of solder mask.

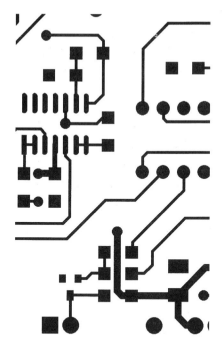

Figure 2-7: Example of a top layer with routing

Figure 2-8: Example of top layer solder mask

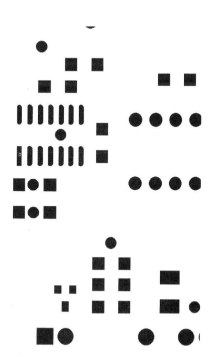

As a side note, there is an ongoing debate as to whether vias should have solder mask over them (called *tenting*) or be tinned. There are good arguments for both sides of the case. My personal preference is not to use solder mask (i.e., tin-plate the vias). This way I can use the via location as a test point when my design doesn't work!

Paste mask: If you are lucky enough to have your circuit boards stuffed at an assembly house, you will need a *paste mask*. Solder paste must be applied to all surface mount component pads before the board is sent through a solder re-flow or infrared oven. To do this, the assembly house needs a paste mask layer. The paste mask layer tells the fabrication house exactly where the paste needs to go. This paste insures that the surface mount pins solder properly to the pads of the circuit board. Through-hole components do not need solder paste. When you make a padstack for a surface mount pin, you should always add a paste mask layer obstacle. This obstacle should have the same size and shape as the top layer pad. You place obstacles on the paste mask layer where you want solder paste. Again, Figure 2-7 shows an example of a section of a circuit board with surface mount pads, through-hole pads, and some routing. Figure 2-9 shows the same area's paste mask layer. Note that the only obstacles on the paste mask layer are the surface mount components.

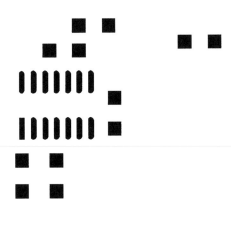

Figure 2-9:
Example of top layer with paste mask

Silkscreen: Silkscreen is a layer that everyone should be familiar with. This layer describes where non-conductive ink is applied to the circuit board after the solder mask. This layer is typically used to outline components, add reference designators, add general text, instructions, etc. Again, Figure 2-7 shows an example of a section of a circuit board with surface mount pads, through-hole pads, and some routing. Figure 2-10 shows the same area's silkscreen.

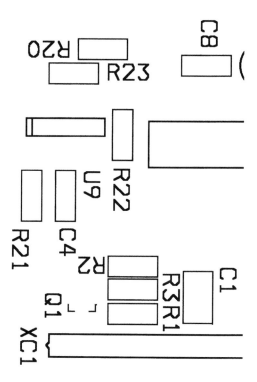

Figure 2-10: Example of top layer with silkscreen

Figure 2-11 shows a completed circuit board manufactured by a professional fabrication house. If you look closely you can see the silkscreen layer, the solder mask layer, and the areas of tin-plating. Notice that the only areas with tin-plating are those that will have components soldered to them. Everywhere else is covered with solder mask.

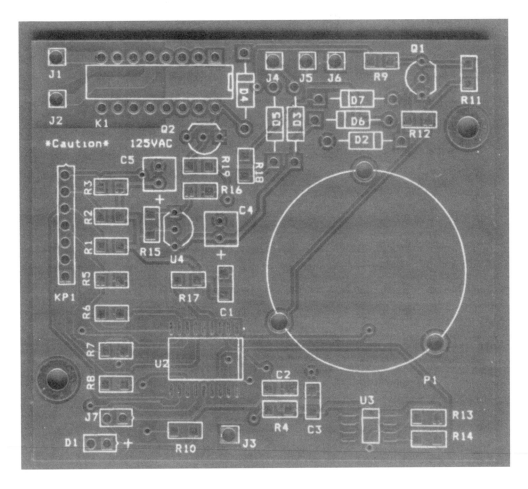

Figure 2-11: Example of a finished printed circuit board

That wraps up our overview of the terminology for printed circuit board placement and routing. Read your documentation that came with your layout tool for the exact details of creating footprints.

Once all the footprints are placed on the circuit board, you must make all the connections. This is where the netlist comes in. The netlist can be used by the layout tool in a few ways. It can be used to insure that the connections made between components in the layout tool are the same connections made in the schematic tool. A design tool called SuperPCB contains a utility that allows you to extract a netlist from a routed design and compare it against the netlist from the schematic tool. The netlist can also be used to drive an automatic routing program. Some layout tools include an automatic router that tries to make all the connections for you. For complicated boards an automatic router can complete as much as 95% to 100% of the connections. SuperPCB is available with an automatic router.

Automatic routing is a great thing to have, but there are a few points that must be considered. You should always "hand route" all critical nets and clock nets first. A *critical net* is any net with very tight timing specifications, high speed, etc. Once these routes are complete you should lock them in place so that the automatic router can not move them. Some automatic routers actually push and shove traces around so that it can fit new traces between them. This type of router is called a *shove-aside* router. It can be very aggravating to have your expertly hand placed traces shoved to the side so a power trace can fit between them! There is more on automatic routing ahead. As I have said before, read the documentation that comes with your routing tool.

Placement and Routing Guidelines

Once you have an electrical design "netlisted" and have settled on a set of design rules suitable to your fabrication process, then it is time to actually place and route your circuit board. Do not underestimate the importance of these steps; careful placement and routing can insure that your circuit board performs to its fullest potential.

The placement and routing of a circuit board is much more of an art than a science. If you ask ten board designers the same question, you will probably get twelve answers, and maybe start a fistfight! Like all art, circuit board placement and routing is best learned by lots of practice. Don't be afraid to make mistakes; I have made plenty. I hope that the following guidelines will help you avoid a few of them.

Automatic Placement and Routing

One question you may be asking is, "Why do I care about placement and routing when my PCB design software can do it automatically?" This is a simple question, but the answer is not so clear. In some cases, for some boards, the automatic placement and routing software may do just fine. For many other boards, it will not. The problem is that you cannot tell ahead of time how your board will turn out. If you don't mind making a few extra boards, then maybe the automatic option is for you.

Automatic placement is particularly troublesome. You should know how your design functions, but your design software only knows how it is connected. There is a big difference. There are many circuits which will work properly when placed properly, and will not work at all when placed poorly. This is particularly true of high speed digital boards, high-performance analog boards, and almost all radio-frequency (RF) boards.

Automatic routing is somewhat better developed than automatic placement. Unfortunately, the routers that have been developed are mainly focused on routing very large digital boards with a lot of connections. They usually will do a poor job on any type of analog or RF board. In addition, most automatic routers tend to generate a lot of vias. This can be a problem for the home hobbyist who has to swage or solder each via individually.

The best thing that I can say about automatic placement and routing software tools is that they have their place, but they should not be used carelessly. You know your design better than any piece of software. If you can place and route the board manually, then you should. The result will be a higher quality circuit board, and you will have thought and learned more both about your design and the design process.

The Influence of Different Electrical Design Types on Layout

An important concept to understand is that different types of circuit boards are usually placed and routed in different ways. We shall consider the following types of circuit boards and provide some useful placement and routing strategies for each (note that these board types are listed in order of increasing difficulty):

✎ *General purpose analog:* This type of board contains analog circuitry (such as op amps and transistors) that operates to only a few megahertz (MHz) of bandwidth. There is usually not much gain, and low noise is not a concern.

✎ *General purpose digital:* This type of board contains digital circuitry (gates, counters, and microcontrollers) that operates up to about 20 MHz. There is usually not much critical timing in the design.

✎ *High-performance analog:* This type of board has analog circuitry that is either higher bandwidth (like video circuitry), or higher gain (like sensor amplifiers), or lower noise, or larger dynamic range (like A/D and D/A converters) than the general purpose analog circuit board.

✎ *High-speed digital:* This type of board has digital circuitry that runs above a speed of 20 MHz. There is often critical timing in the design.

✎ *Radio frequency (RF):* This type of board has special circuitry designed to operate at very high frequencies (above 20 MHz), and often with very low noise and high dynamic range requirements. The circuit board itself plays a much larger role in the performance of an RF design.

Of course most real-world designs are a mix of the above circuit board types. In those cases you must artfully blend the placement and routing strategies to produce the best circuit board. Remember, there is no "right" way to place and route a circuit board, but there are plenty of ways that will result in a circuit board which never works as well as it should!

General Placement Considerations

Before discussing the specific board types listed above, a little time should be spent on general placement considerations. A good placement almost always lead to good routing, and thus to a good board, but what makes a "good" placement? Well, a good placement is one that leads to a good routing. Of course, this is circular logic, but it's true anyway!

One rule of thumb is to place components close to each other while trying to keep the routing process in mind. If your PCB design software supports *ratsnests* (or guides) to show where the connections will be, use them. They clearly show how the board will be routed as you are placing the components. As you place components, turn and flip each one around to make the guide lines as straight and as short as possible.

Another rule of thumb is to place components on the circuit board just like they are drawn on the schematic. That means that components that are directly connected to each other will be placed closely together. It is usually easiest to place the big components first (like microcontrollers, op amps, etc.) and then place all of the little ones that connect to them (resistors, capacitors, etc.) around them.

It is almost always best to completely place all components on the circuit board before beginning any routing. If your PCB design software has placement design checks, run them after the board is placed. This will inform you of violations such as components placed too close to each other (or on top of each other!), or components placed too close to the edge of the circuit board.

Finally, try to leave at least 30 mils between components, and 50 mils between a component and the edge of the circuit board. It is possible to place components closer together, but this becomes an issue when you try to assemble the board. If the components are too close, the solder may flow from pad to pad and generate short circuits.

Special Placement Consideration: Decoupling Capacitors

There is one type of component that appears in almost every type of design: the *decoupling capacitor*. The word "decoupling" really means "noise-absorbing." These capacitors are located on the power input pins of nearly every type of active device. Their purpose in life is to absorb (or more correctly, shunt to ground) electrical noise which is generated in the active device. This prevents an active device on a circuit board from spewing noise onto the board's power supply, and disrupting other devices. It is extremely important to have decoupling capacitors (usually 0.1 µF ceramic) on all types of digital circuits and on most analog ones as well. Decouplers must be closely placed adjacent to the device power pin, or they are useless. This is because if you place them far away, the trace which connects the decoupler to the power pin acts as an inductor, and makes the capacitor "disappear" from the circuit.

General Routing Considerations

There are no rules for determining "good" vs. "bad" board routing, but there are many factors to consider as you route your board. In general, the shortest routes are the best. This is because long routes tend to generate more electrical noise, or pick up more electrical noise. You can think of each trace or route on the circuit board as a small antenna. If there is a lot of signal energy present, the trace will act like a transmitting antenna. If it is connected to a sensitive device, it will be a receiving antenna. If the trace is connected to RF circuitry, it may become both a transmitting and a receiving antenna! One way to minimize these effects is to make all of the routes as short as possible. In fact, many PCB routing tools use route length as a figure of merit to determine how well their auto-router is working.

One area of special concern when routing a board is "cross-talk" or "cross-coupling" between routes. This means that the signal energy on one route couples through the circuit board (or the air!) to another route. The longer the traces run next to each other, the greater the cross-talk will be. This occurs quite frequently on high-speed digital circuitry, as well as on high-performance analog and RF boards. The effects can be devastating. It is not uncommon to have a circuit board that is completely non-functional due to cross-talk. When this happens, the only solution is to physically cut the affected traces and run wires instead. Obviously, you want to avoid this at all costs. Once again, there is no free lunch and no way to completely eliminate cross-talk from your board design. Instead, you must be vigilant as you route your board.

Pay attention to the signals you are routing, because some are more likely to generate interference than others. For example, a high-speed digital clock signal is going to spew noise everywhere. Don't route it next to the input to your high-gain op amp. To do so only invites disaster! One thing to look out for is long traces routed next to each other or on top of each other (on a multi-layer board). Auto-routers are fiendishly good at making these structures. Two long traces routed next to each other will generate a lot of coupling. It's OK to route like that if the signals "don't mind" being coupled, like two data bits of a parallel address bus. It's not OK to route like that if one of the signals is particularly sensitive, like the input to an amplifier.

Another thing to consider while routing your design is the number of vias you use. If you are having your boards professionally made with plated-through vias and holes, then you probably don't mind having a lot of vias. (Of course the fabrication house will usually charge you more if you have a lot of vias!) If, however, you have to build each via yourself (either with wire and solder or with swaged-in rivets), then you will want to minimize the number of vias you use. Again, most PCB routing tools offer via minimization options. Finally, never forget that vias can couple signal noise just like traces can. It doesn't do any good to carefully route a sensitive trace right next to a "noisy" via.

You can see that a "good" routing job requires you to be aware of every signal on your board, what it does, and how sensitive it might be to interference. A good rule of thumb to remember is that interference sensitivity (also called susceptibility) increases with frequency, impedance, and gain. This means that a circuit that operates at high frequency (20 MHz or above) is usually more susceptible than one which operates a low frequency (20 kHz). Ditto for impedance and gain. If you see a circuit with a lot of small capacitors (like 33 pF or smaller) and inductors (like 100 mH or smaller), it is a good guess that both the frequency and impedance are high, so board routing will be critical. If you know that a circuit has high gain (like some op amps, or most IF and RF amplifiers), routing will be important. If you route the output of a high gain circuit near the input and the signals cross couple, your amplifier will become an oscillator!

Special Routing Consideration: Ground

Ground is the most important and most often ignored signal on all circuit boards. More good circuit board designs have been rendered useless due to bad grounding than for any other reason. Many designers place and route their entire board, and route the ground signal last. This may be okay on some boards, but it is devastating on many others. Almost all active devices require rock-solid grounding to meet their manufacturer's specifications, and often even to function correctly. As devices operate faster and at lower voltages, grounding becomes more and more important.

The reason that grounding can so easily become a problem is that every signal on the board eventually flows into ground—every current, every voltage, everything! In a well-grounded board, the ground signal is a solid reference for every signal on the board, no matter the signal's frequency, size, or location on the board. Unfortunately, the reverse is also true. A poorly grounded board actually has no fixed reference point. It may be that the DC value of "ground" varies in different locations of the board, or even contains an AC, or high-frequency component.

So what makes a ground "solid"? The answer is the *impedance* of the ground. If you route the ground net with very narrow traces, it will have higher resistance than if you route it with wide traces. If you then have large currents flowing though the ground routes (remember, every current eventually flows through ground), the high resistance will cause a voltage drop (Ohm's Law: $V = IR$). This voltage drop makes ground have different voltages at different places on the board. What does this mean? You should route ground with wide traces rather than narrow ones. Wide traces also have another beneficial effect. At high frequencies, a wide trace is less inductive than a narrow one. This means that the high frequency AC resistance of your ground is lower. The goal here is to create a ground signal that has the lowest resistance at any frequency or board location.

There is one thing even better than wide ground traces: *ground planes*. A ground plane is just one giant area of copper on your board that becomes your ground. In fact, if you can make one whole side of your board into ground, then you have the best possible solution. This solves your ground routing problems too. If you have a ground side to your board, a good ground is just a via or through-hole component away.

Even if you can't dedicate one entire layer of your board to ground, you can come close with large ground areas on both top and bottom of your board. Most PCB design tools allow you to have copper areas (often called copper pours or area fills). These can be "poured" around existing routing at the end of your board design. Many layout tools provide adequate clearance to avoid "short-outs," but will still connect all of the ground pins to the copper area. I usually put a single copper area the entire size of the board on both sides. This way the areas on each side will connect together by any ground vias or through-hole pins. This copper area technique has a few hidden benefits as well. The abundance of ground on your board will tend to provide built-in shielding from external interference, as well as help quiet any noisy routes and reduce cross coupling on the board. In addition, the ground provides a good safe path for static electricity, thus increasing the reliability of your finished board. Finally, you use less developer when you make the board, because you etch away less copper (unfortunately you do use up a lot more tin plating solution). See Figure 2-12 for a sample of a ground fill used on a circuit board. Note how the ground fill covers all unused area of the board, but stays clear of traces. The ground fill is solid copper.

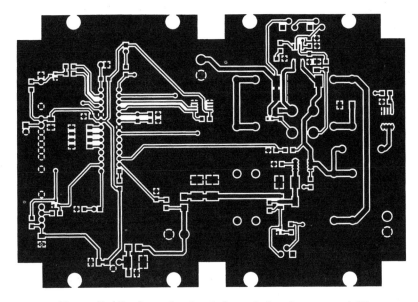

Figure 2-12: Sample circuit board showing ground fill

Look carefully at the board and find some grounded pads—you will see that they are connected to the ground fill by thin traces. These traces are called *thermal reliefs*, and allow the board assembler to heat a grounded circuit pad without heating the entire ground fill. Note also that the ground fill is removed in certain areas of the board. This is because, in those areas, the fill could not be connected to ground on that layer. If the ground fill were there, it would be an isolated island of metal, electrically connected to nothing on the board. This could be a severe board problem, since the "floating" metal area could act as a large antenna or coupler, and cause noise or interference problems to the other circuits on the board.

Special Routing Consideration: Power

Power distribution can be as nearly as important as circuit board grounding, and most of the above comments apply equally to ground or power signals. After all, the power signals on your board can be thought of as ground signals with different DC voltages. They still need to have low resistance, both DC and AC. You should route power distribution networks with fat traces, power planes, and copper areas. When used for power distribution, narrow traces become fuses, and quickly burn out! Before routing any power signals, spend a few minutes and consider the current carrying capacity of the traces you intend to use. Luckily for you, the next section deals with this exact topic.

DC Line Resistance and Current Carrying Capacity

An issue commonly overlooked during printed circuit board design concerns the resistance of the circuit board traces. Circuit board traces can be thought of as really just flat wires. And just like a wire there is a limit on how much current you can pass through them before resistive losses create excessive voltage drop and heat buildup. Let's take a look at just how much current you can pump through a circuit board trace. Note that all the information below relates to DC.

Resistance is a function of the resistivity of the conducting material and its geometry, as illustrated in Figure 2-13:

Resistance, $R = \rho \ell \, / \, A$

ρ = resistivity of copper $\qquad = 17.24 \times 10^{-9}\ \Omega \cdot m$

$\qquad\qquad\qquad\qquad\qquad\qquad = 6.787 \times 10^{-4}\ \Omega \cdot mil$

ℓ = length of trace

A = cross-sectional area of trace

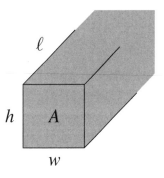

Figure 2-13: Conductor cross-section

You should note that the resistivity of copper increases linearly with temperature. Specifically

$$\rho = \rho_{rt} [1 + \alpha(T - T_{rt})]$$

$$\rho_{rt} = 6.787 \times 10^{-4} \ \Omega \cdot \text{mil}$$

$$T_{rt} = 20°C$$

What does all this mean? As the temperature rises, so does the resistance of a conductor. Table 2-1 lists the resistivity of a copper conductor at various temperatures and the third column lists its resistance per unit of a 5-mil inner layer trace using 0.5-oz foil.

Table 2-1

Temperature (°C)	Resistivity ($\Omega \cdot$ mil)	Resistance (Ω/inch)
-40	5.19×10^{-4}	0.44
-20	5.72×10^{-4}	0.49
0	6.25×10^{-4}	0.53
20	6.79×10^{-4}	0.58
40	7.32×10^{-4}	0.63
60	7.85×10^{-4}	0.67
80	8.39×10^{-4}	0.72
100	8.92×10^{-4}	0.76

Now to put this into terms you can use. Figure 2-14 shows resistance of a conductor versus the width of the conductor for three common copper weights. When you buy circuit board material, or have a board fabricated by a professional board house, you specify the copper weight (a measure of its thickness) in ounces. Typical values are 0.5 ounce, 1 ounce, and 2 ounces. (The circuit boards used in this book are 1 ounce.)

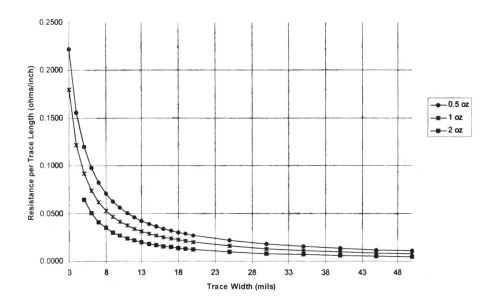

Figure 2-14: Conductor resistance versus width

You should note that the resistance and resistivity vary slightly based on whether the conductor exists on an inner or outer layer. Since you can only produce outer conductors with the process outlined in this book, Figure 2-14 applies only to outer layers. Inner layer characteristics are not really much different.

The IPC-D-275 standard has charts that show the relationship of the current carrying capacity at various temperature rises with respect to the cross-sectional areas. Charts are provided for both inner and outer conductors. Using these charts, a designer would be able to estimate the conductor thickness and width necessary to support a given current value and a predicted temperature rise of the system.

After all that, an example is in order. Let's say that we have a 4-inch long, 18-mil width conductor made from 1-ounce copper. If there were 50 milliamps of current flowing through the trace, how much of a voltage drop would there be through the trace? By examining the curves in Figure 2-14, we see that for an 18-mil conductor made from 1-ounce copper the resistance per inch is approximately 0.025 ohms/inch. Therefore, the total trace resistance would be:

(0.025 ohms/inch * 4 inches) = 0.1 ohms

Now, using Ohm's Law, the voltage drop would be:

0.1 ohms * .050 amps = 0.005 volts

In other words, it is negligible. For normal CMOS/TTL loads you usually do not have to concern yourself with the resistance of the conductors. When you start driving large currents, it is a good idea to check the resistance of the conductor in order to size it accordingly.

CHAPTER 3

Placement and Routing Considerations for Various Circuit Designs

Placement and Routing Considerations for General-purpose Analog Circuit Boards

A "general-purpose" analog circuit board is defined here as one that uses common analog circuitry (op amps, transistors, etc.) that operates at only a few megahertz (MHz) of bandwidth. This type of circuit is fairly forgiving of circuit board placement and routing oversights, but there are a few guidelines still to be followed. Proper ground and power distribution (as described in Chapter Two) is important, as are decoupling capacitors. Beyond that, you should not have many problems.

Placement and Routing Considerations for General-purpose Digital Circuit Boards

"General-purpose" digital circuit boards contain digital circuitry (gates, counters, microcontrollers, etc.) that operates up to about 20 MHz. There is usually not much critical timing in the design. Once again, this board type is fairly forgiving of placement or routing oversights. Since the circuitry on the board is digital, however, there are a few other things to think about when placing and routing your board.

Most modern digital circuitry is CMOS (complementary metal oxide semiconductor) of one type or other, rather than the older TTL (transistor-transistor logic). CMOS logic consumes much less power and switches faster than TTL. The increased speed comes with a price, however; the price is noise! The extremely fast transitions on CMOS logic outputs—1 nanosecond (ns) or less—produce much more power supply noise and crosstalk than TTL logic.

To prevent CMOS logic from corrupting your power supply, be sure to use appropriate decoupling capacitors. As a minimum, you should have one capacitor (0.1µF ceramic is typical) per CMOS chip. Place the capacitor close to the chip's power pin, as described in Chapter Two. Devices with more than one power pin need more than one decoupling capacitor; one decoupling capacitor per power pin is sufficient.

Controlling crosstalk on a general purpose digital circuit board is usually not too difficult. Be sure not to route outputs next to inputs, especially if the inputs are to be used when the outputs are switching. This will prevent the sharp transition of the output from coupling onto the input line at just the wrong time. Of course, it is impossible to route the board without getting inputs and outputs close together somewhere; just try to avoid long lines where the coupling might be worse.

One final note about general-purpose digital circuitry involves crystal oscillators. Most microcontrollers use a crystal oscillator as a timing reference. The crystal is used in a positive feedback loop around an amplifier in the microcontroller. The positive feedback makes the amplifier unstable, so it oscillates. If everything goes right, the amplifier oscillates at exactly the frequency of the crystal. Who cares, you ask? Well, you do! If you are faced with a board design containing a crystal oscillator, you must be careful both with placement and with routing. Most crystal circuits require a small capacitor (typically a 15-pF ceramic) to ground on each side of the crystal. In addition, some circuits have a resistor in series with the feedback loop. To insure that your oscillator will work, place it first (right after the microcontroller). Try to place the crystal and resistor in a small loop, right next to the pins of the microcontroller. Place the startup capacitors close by. Route the board with the shortest routes possible, and make sure the ground connections on the capacitors are very short. See Figure 3-1 for a correctly placed and routed crystal oscillator. Notice how X1 (the crystal) is in a small loop of traces starting and ending at U1 (the microcontroller). R2 (the series resistor) is in the loop, while C1 and C2 (startup capacitors) are connected from the two loop nodes to a ground plane (not shown). Notice how the parts are placed as close together as possible, and how the traces form a small, tight loop.

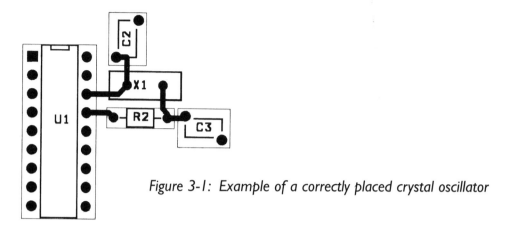

Figure 3-1: Example of a correctly placed crystal oscillator

Placement and Routing Considerations for High-performance Analog Circuit Boards

"High-performance" analog circuit boards contain analog circuitry that is either higher in bandwidth (like video circuitry), higher in gain (like sensor amplifiers), lower in noise, or have larger dynamic range (like A/D and D/A converters) than a general-purpose analog circuit board. No matter what type of circuitry is on your board, there are a few useful techniques for placing and routing high-end analog circuitry. By this point, it should go without saying that good grounding, power distribution, and decoupling techniques must be used.

The active devices (op amps and transistors) on a high-performance board tend to have both more gain and operate at higher frequencies. This can yield significant problems if a few guidelines are not followed:

- Be sure that amplifier outputs and inputs are separated.

- Be sure that op amp feedback loops are physically small. Place feedback resistors and capacitors as close as possible to the op amp pins.

- Remember to always put a small capacitor (100-pF ceramic) in parallel with the op amp's negative feedback resistor to reduce gain at high frequencies, where oscillations usually happen.

- If you have a large amount of gain from multiple devices in series, make sure that the output device is far from the input device.

✍ If you have a large amount of gain from multiple devices in series, make sure that the input and output devices are not in the same package. In this case, it is often better to use two dual op amps instead of one quad op amp.

In addition to these guidelines, the performance of analog circuitry can be improved by a technique called *ground rings*. A ground ring is used to isolate a section of circuitry from the rest of the board. It is most useful for high-end op amps. See Figure 3-2 for an example of a ground ring. See how the ground ring surrounds the sensitive input pins of the amplifier. It connects both to the ground pins of the amplifier and to the ground plane of the board (not shown). Ground rings should be used on low noise or high impedance points in the circuit.

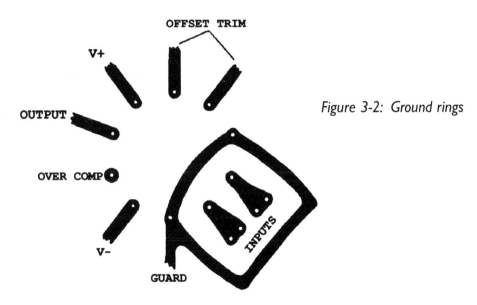

Figure 3-2: Ground rings

Another important routing technique involves differential signals. These signals are often used to improve a system's dynamic range and noise immunity. A differential analog signal is actually a pair of signal lines with opposite DC values. In a typical differential signal application, a pair of differential signal lines is used to move a signal from a driver to a receiver. Note that these parts can be quite a distance from one another. The differential driver (usually built from an op amp) drives each signal line in opposite directions. At the other end,

the differential receiver (also usually an op amp) subtracts the two signal lines from each other to produce the output signal. When routing these types of signals, it is important that the two lines be routed completely parallel and close to each other. The reason for this is so that any external noise will be coupled equally onto each line, and thus subtracted out at the receiver. This is one of the only times when long parallel routes can actually work to your advantage.

Placement and Routing Considerations for High-speed Digital Circuit Boards

This type of board has digital circuitry that runs above 20 MHz. Common components in this type of design include microprocessors or digital signal processors, static or dynamic RAM, flash memory, high-speed programmable logic, and complex mixed signal processors. There is often critical timing in the design, where signals must arrive within one nanosecond (ns) of each other. These designs also usually have large (more than eight bits) address and/or data busses which must be connected between chips.

The high-speed operation of these large busses creates tremendous electrical noise. When 16 or 32 CMOS lines change state simultaneously, a large amount of energy is required from the power supply. The supply cannot provide the power fast enough, because of the inherent inductance in even the best-designed distribution system. For this reason, power supply decoupling is even more important. Every pin which connects to the power supply must have a decoupling capacitor (0.1 μF ceramic is typical). This means that, for a microprocessor or DSP chip, there may be as many as 20 decouplers! In addition to the 0.1 μF decouplers on each power pin, you should provide a "bulk" decoupling capacitor for each large digital device. This is usually a larger capacitance (10 μF tantalum is typical) to provide an energy store for this hungry digital chip!

Be careful when you route the decoupling capacitors. In order to prevent the noisy chip from corrupting your power supply, you must make sure that the capacitor is located between the chip and the supply. Figure 3-3 shows a poorly placed and routed decoupling capacitor, while Figure 3-4 shows a correct decoupling design.

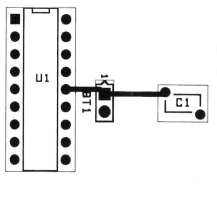

Figure 3-3: Incorrectly placed and routed decoupling capacitor

Figure 3-4: Correctly placed and routed decoupling capacitor

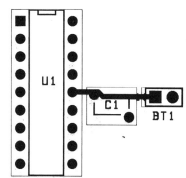

In both Figure 3-3 and Figure 3-4, U1 is the active device. It connects to the power supply on pin 14, and the power is provided by connector BT1. C1 is the decoupling capacitor. Notice how in Figure 3-3 the power flows in from BT1 directly into U1, while C1 is way off to the right. In fact, as far as high-frequency noise is concerned, C1 is effectively out of the circuit. The long trace length between U1 pin 14 and C1 will contribute enough stray (*parasitic*) inductance such that the effective value of C1 is greatly reduced. Now consider the circuit in Figure 3-4. In this case, the power must flow in from BT1, through the input pin of C1, and into U1 pin 14. In this case, C1 will effectively decouple any high frequency noise produced by U1. Note that the idea here still applies no matter how the power is applied to the active devices. In the examples, I used connector BT1, but you might use a trace, or a via, from a backside or inner power layer. No matter how the power gets to the chip, it should have to go through the decoupling capacitor and from there through a short trace directly to the device power pin. You cannot decouple a circuit board too well.

When routing the high-speed digital circuit board, keep all routes as short as possible. If the route lengths get too long, the fast logic transitions may cause signals to bounce back and forth up and down the route, causing *ringing* on the desired signal. This ringing is also called a *transmission line* effect. The signal ringing can be reduced by terminating the routes with a resistor. A *shunt termination* is often used. This is usually a "pullup" (to the power supply) or a "pulldown" (to ground) resistor of approximately 10 kilohms. The shunt termination is placed at the destination of the route. Keep in mind that the shunt termination will dissipate power and load the logic gate that is driving the line.

In some cases, a shunt termination may not be enough to reduce the ringing to acceptable levels. This is often true of high-speed clock lines. In these cases a *series termination* will usually reduce transmission line effects to tolerable levels. The series termination is usually a small resistor (150 Ω typically) placed at the source of the signal.

High-speed digital clocks are among the most difficult signals to route. A common problem is how to distribute a clock from some type of source (an oscillator or gate output) to several different inputs across the board. There are basically two ways to do this: a "starburst" or a "daisy-chain" route. The *starburst* route places the clock source in the center of a "star," with each clock load at the end of a route. The problem here is that each arm of the star is a transmission line. If these lines aren't terminated correctly, the reflections and ringing will all combine at the center of the star, and can ruin the clock signal. If you do terminate each end of the star, the clock source may not have enough power to drive all of the terminations.

The other solution is to route the clock signal in a *daisy chain*. This is a single, long route from the source, through each load one by one. The line should be terminated at the end to avoid ringing. This is a more difficult way to route the clock, but will often give superior results.

Placement and Routing Considerations for RF Circuit Boards

Radio frequency (RF) circuit boards are generally considered the most difficult boards to place and route. This is due to the characteristics of the circuits and signals found in typical radio designs. Signals on an RF board range in frequency from zero (DC) to tens of gigahertz (10,000,000,000 cycles per second!). In addition, the *amplitude* (size) of the signals ranges from tens of nanovolts (0.00000001 volt) to tens of volts. Circuits commonly used in RF designs often have very high impedance. This, combined with the high gain commonly found in radio designs (a gain of 100 dB, which is ten billion, is not unusual!), greatly increases the circuits' susceptibility to interference or pickup. Finally, the component values used in RF circuits are often very small. Capacitors as small as 0.1 pF (0.0000000000001 farad) and inductors as small as 10 nH (0.00000001 henry) are commonly used. All of these factors combine to make an RF circuit board a true placement and routing challenge. A good RF board designer must keep each of these difficulties in mind as he or she places and routes the board. Note that I said "he or she," not "it." There is no place for automatic placement or routing software in the design of RF circuit boards. If you think you will save time by using these tools, then think again. You will only force yourself to completely redo the board (by hand) after it doesn't work!

RF Circuit Board Special Placement Considerations

It has been said that the three most important factors in real estate are "location, location, and location." That may or may not be true, but it is definitely true that the three best ways to design an RF circuit board are "placement, placement, and placement"! No amount of clever routing or grounding will make an incorrectly placed RF board function well.

When placing the board, start at one end of the schematic (like a radio receiver's antenna) and work towards the other end (such as the speaker). It is often best to work from high frequency to low; in this way you place the most critical components first. As you go, place active devices (ICs or transistors) first, then place any passives that are directly in the signal path. Next place any decoupling capacitors or choke inductors. Ignore (for now) any other components. This way you can place the RF signal chain first and fill in any support or control components later.

As you place the signal chain, think about the function and frequency of each device. Do not place active devices near each other that have high gain, or that operate in the same frequency band. For example, if one chip is a receive IF strip with a high gain (maybe 80 dB) at 45 MHz, and another chip is a transmit amplifier which will generate a midlevel signal (maybe +10 dBm) at 90 MHz, do not place them together. It is a good bet that your IF strip will still have plenty of gain at 90 MHz, and proximity to the transmit amplifier will cause it to overload and saturate. If the transmit chip was instead operating at 900 MHz, then placing these components together might not be a problem, since the presence of a 900 MHz interfering signal should have no effect on the IF strip at 45 MHz. I find that it is best to break up the board into functional areas. I place the transmitter components in one area, the receiver in another, and the oscillators in yet another.

The signal chain should be placed in as straight a line as possible. Try not to generate any crossovers, especially in high gain IF sections. If you have to bend or cross over the signal chain, try to do it such that the points that overlap (or come close to each other) are at greatly different frequencies. Try to keep the signal chain on one side of the board. If that isn't possible, change sides at only a minimum number of places, and try to separate the circuitry on each side from each other. Remember that two components on opposite sides of the board are actually very close, and will interfere with each other. Place each component as close as possible to the previous and the next, leaving room only for the necessary routing and ground vias, as well as the support components you will put it later. The closeness of the components will greatly reduce the effects of unwanted (or parasitic) capacitance and inductance on the board.

The placement of decoupling components, as always, is very important. It is common in RF designs to have a device pin bypassed with two or more decouplers, often of differing value. Place the smallest value component closest to the pin, with the larger ones further away. This will give the maximum performance from the smaller components at higher frequencies. Also, give special thought to any oscillator circuits on your board. In addition to generating a lot of potential interference energy, oscillators are extremely susceptible to external interference.

Remember as you place the board that every component, especially active devices, small capacitors, any inductors, and any resonant circuit can act as both a transmitting and receiving antenna. All of these little "antennas" on your board will try to communicate with each other and will seek to change the operation of your circuitry, usually for the worse. A little thought at the earliest stage of placement will go a long way to eliminate these interference problems. Placing an RF circuit board is at best an acquired skill, and at worst it is black magic. Take your time and think about what you are doing. If you completely understand the function of every component in the design, then you will be able to place each one in the best possible spot.

RF Circuit Board Special Routing Considerations

Once your RF board is placed, it is time to think about routing. I said "think," not "start routing"! If you just start connecting things, you can easily waste all of the time you invested in your careful placement. So calm down, and think about what you are going to do, and then do it, and then think about it again!

The first thing to think about in routing your RF design is *grounding*. All of the comments for the other board types apply here, and more. Ground is the single most important feature of the RF board. A poorly grounded RF board will soon be scrap. As always, the best ground is the most ground. If you have the luxury of a complete ground plane, then by all means use it. In fact, there are many cases where two or more ground planes may be required (see the section on stripline later in this chapter). If, however, you are making the boards yourself and are limited to two layers, then you must fill each layer with as much ground as possible. In addition, place as many vias as possible to tie the ground areas on each side of the board together. These should be placed no further than every 0.5 inch. Finally, it is a good idea to leave enough room to wrap metal (copper) tape around the edge of the board and solder it to the ground areas on each side. This will provide the most solid RF ground possible. See Figure 3-5 for an example of a well-grounded RF board. Notice in the figure how any ground pins are connected through very short traces to vias. These vias then connect to a ground plane (not shown) on the other side of the board. Also note how the large filter component (in the center) and the two connectors (on the bottom) have copper fill areas to provide the most solid RF ground possible.

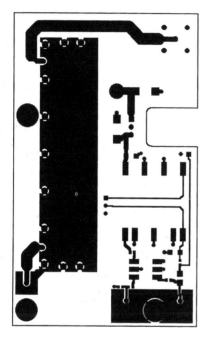

Figure 3-5: A well-grounded RF board

Once you have a grounding concept, I recommend you route the grounds for all of your RF components. These will be the first routes on your board, so you can make them right. Keep each route as short as possible, as measured from the pin of the device you are grounding to the ground plane or area. This will minimize parasitic ground inductance. It is only necessary to ground the RF devices now; leave all other routing for later.

Now it is time to route the signal chain. If you have a good placement, this will be an easy job. The signal chain is the path through the circuit that the primary RF signals follow. For example, a receiver's signal chain would start at the antenna input and flow through to where the signal is no longer considered RF, usually at audio frequencies. The signal chain routing should consist of a series of very short connections between components, with a minimum of vias, crazy turns, and loops. If you start seeing these features, then your placement isn't good enough yet. Rip up *all* of the routing and revisit your placement. Do not hesitate to delete routing that doesn't look right. A better placement will almost always present itself.

Again, remember not to run amplifier inputs and outputs near each other, especially in high gain circuits. In a more general way, you should not route high-level signals near low-level ones; the large signal will couple easily onto the small one. Of course, it would be best to keep all RF signals far apart from each other; however, on a complex or small board, this will not be possible. In that case, try to keep signals of similar frequency furthest from each other while letting ones that are of greatly different frequency come closer. For example, suppose you are forced to route an 800-MHz signal and a 455-kHz signal closer than you would like, and the signals couple onto each other somewhat. The frequency response of the devices connected to the routes will tend to reject the effects of the interference. An 800-MHz amplifier will likely not respond to the 455-kHz interference coupled into it, and a 455-kHz amplifier will certainly not respond any 800-MHz signal energy. Note that the 455-kHz energy which couples onto the 800-MHz signal could cause spurious *sidebands*, which is a type of RF interference.

After the signal path is completely routed, it is time to connect the routes for the miscellaneous and support components. Don't think these routes are any less important than the signal path routes; a careless route of a noncritical signal can easily corrupt a more critical one. Be careful not to put any "miscellaneous" traces between "signal path" signals you were trying to keep apart. Any trace, or even a via, in the wrong place will provide a mechanism for critical signals to interfere with each other. Be especially wary of control lines to or from digital logic—they are often noisy by nature and their high impedance allows them to act as excellent RF couplers.

Remember, routing an RF board is a challenge. Don't rush, and never be afraid to rip up your routes and start again. It is often best to work in short sessions, to keep your concentration level up. If you run into a problem that you cannot resolve by moving your existing routes around, then your placement isn't right yet. In that case, you *must* rip up *all* routes in the area, fix the placement, and then start the routing again. Don't be tempted to ignore any routing problem you know about, no matter how insignificant. A single trace, via, or component in the wrong place has put many an RF board into the scrap heap, and caused many more board revisions. It's much more work to do the board over than to do it right the first time!

Special RF Routing Techniques: Microstrip

When routing high-frequency RF signals, the traces themselves become an important part of the circuit. Remember the high-speed digital circuit boards discussed above? The traces on those boards are sometimes called *transmission lines*, because of the reflections that happen at their ends. The solution there was to terminate the lines with resistors to try to reduce or eliminate these reflections. A different solution is often used on RF transmission lines: the lines themselves are designed to match the terminating impedance inherent in the RF circuitry. Two ways to achieve this are with the use of *microstrip* or *stripline*. These are types of *controlled-impedance* traces. They are easy to design and build, even for the home hobbyist. We will discuss microstrip first and then stripline in the next section.

Which RF traces should be treated as transmission lines? In general, all high frequency or long traces should be considered transmission lines. But how high is high or how long is long? A rule of thumb to use is that any trace that is longer than 1/16 of a *wavelength* will act as a transmission line. To compute a signal's wavelength, use the formula:

$$\lambda = \frac{11232}{F}$$

where λ is the wavelength (in inches) and F is the signal's frequency (in MHz). This equation includes an approximation to account for the fact that your signal will travel slightly slower on a circuit board than it would in free space.

For example, a 915-MHz signal has a wavelength of about twelve inches, so a transmission line would be any trace longer than about 0.8 inch. Note that this doesn't mean that you should carefully route a two-inch trace as a transmission line while carelessly routing a 0.7-inch trace. Once you have one controlled impedance transmission line on your board, it is just as easy to make all of your RF traces into transmission lines. See Table 3-1 for a list of frequencies, wavelengths, and transmission line lengths. The table also includes the length of a quarter-wave dipole antenna for each frequency. A quarter-wave dipole is a very common type of antenna; it consists of a quarter-wave conductor mounted over a ground plane. You can see that, at high frequencies, circuit board traces can easily become efficient antennas!

Table 3-1: Signal Frequencies, Wavelengths, and Transmission Line Lengths

Signal Frequency	Wavelength		Minimum Transmission Line Length	Dipole Antenna Length
(MHz)	(inches)	(feet)	(inches)	(inches)
1	11232.0	936.0	702.0	2808.0
10	1123.2	93.6	70.2	280.8
20	561.6	46.8	35.1	140.4
50	224.6	18.7	14.0	56.2
100	112.3	9.4	7.0	28.1
500	22.5	1.9	1.4	5.6
1000	11.2	0.9	0.7	2.8

Once you have identified which traces on your board are to be treated as transmission lines, then it is time to decide on the type of transmission line to use. There are many different types to choose from. You are already very familiar with the most common type of transmission line, the coaxial cable. It certainly works well, as millions of cable TV viewer will attest to. Unfortunately, coaxial transmission lines are difficult to fabricate on circuit boards. As previously noted, the two most common types of transmission line used on circuit boards are microstrip and stripline.

Microstrip is the most common type of transmission line. It simply consists of a copper trace of a known width a certain distance above an infinite ground plane. A dielectric (insulating) material is between the trace and the ground plane, and nothing is above the trace. Figure 3-6 shows a cross-section drawing of a microstrip.

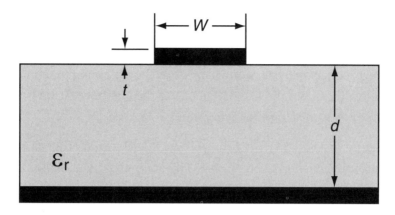

Figure 3-6: Cross-section of a microstrip

Note that any trace on a printed circuit board can be viewed as a microstrip. In fact, all circuit board traces are microstrips, but only some are designed that way. The complete analysis of a microstrip is a bit complex but a few approximations will help. The first approximation is that the ground plane is finite in size, since it cannot really be infinite, but it must be much larger in all directions than the transmission line. Another approximation concerns the free space above the trace. Most circuit boards are used in an enclosure of some type; if a microstrip is used on the circuit board then any metal cover must be kept far above the traces. This will keep the electric field lines above the microstrip from becoming distorted by the presence of a cover.

These approximations allow one to write a simple formula for the characteristic impedance of the microstrip. This equation is:

$$Z_0 = \frac{87 \ln\left[\dfrac{5.98d}{(0.8w+t)}\right]}{\sqrt{\varepsilon_r + 1.41}}$$

where w is the width of the microstrip trace (in mils, which are 0.001 of an inch), d is the distance to the ground plane (again, in mils), t is the thickness of the trace (in mils) and ε_r is the relative *dielectric constant* of the dielectric material used in the circuit board. The result is Z_0, which is the characteristic impedance of the transmission line in ohms. The term *characteristic impedance* is simply a way of describing the effects the transmission line will have on the signals it carries. For the purposes of nearly all circuit boards, your goal as a board designer is to produce transmission lines of a known characteristic impedance. This will allow the circuit board to "match" to the components on it, resulting in a minimum of transmission line reflections.

What is the dielectric constant (ε_r), you ask? It is a material property of the circuit board you have chosen to use, and it is usually outside of the board designer's control. Nearly all (99%) of circuit boards are built on "FR-4" woven fiberglass circuit boards, including the ones used in this manual. Exotic high frequency, high power, or low-loss circuit boards are sometimes built on more advanced (and much more expensive) materials. Table 3-2 lists material properties for common circuit board materials. Air is listed for reference only (it's very hard to build circuit boards in air!).

Table 3-2: Properties for Circuit Board Materials

Circuit Board Material	Dielectric Constant	Loss Tangent
FR-4	4.5	0.02
Ceramic (a.k.a. alumina)	9.8	0.0001
Teflon (a.k.a. duriod)	2.1	0.0003
Air	1	0

In addition to dielectric constant, the *loss tangent* is also given for each circuit board type. This is related to the loss an RF signal will experience when it passes through circuit boards made with each dielectric. A lower loss tangent means less loss. Notice how the least expensive material (FR-4) has the highest loss tangent, while the more exotic materials have lower loss tangents.

The process of computing the loss on a transmission line from the loss tangent is even more complex than the computations for characteristic impedance, as it involves the frequency of the signal. Table 3-3 lists a few representative losses of different types of microstrip traces. Notice how the microstrip width is the same for any signal frequency; only the RF loss rate changes with frequency.

You can see that the RF loss of FR-4 greatly increases at signal frequencies above 1000 MHz. The advanced materials are used exclusively for these high frequencies.

Table 3-3: RF Losses for Various Microstrips

Board Material (1/16 inch thick)	50Ω Microstrip Width (mils)	Signal Frequency (MHz)	RF Loss Rate (dB/inch)
FR-4	115	1	0
FR-4	115	1000	0.09
FR-4	115	10000	0.9
Ceramic	71	10000	0.08
Teflon	206	10000	0.04

The RF loss rate is specified in units of decibels per inch (dB/inch). To compute the actual loss of a stripline, multiply the loss rate by the length of the microstrip. While these losses may seem small, remember that a total loss of only 3 decibels (dB) means that half an RF signal's power is lost in the microstrip, and only the remaining half may be delivered to the load.

Once you have determined what board material and dielectric thickness you are going to use, it is time to compute the microstrip line width to use to design a microstrip of a given characteristic impedance. The impedance you are shooting for will nearly always be 50Ω; this has become the industry standard for RF controlled impedance circuitry. The only significant exception would be for cable television (CATV) designs, which are typically 75Ω. If you re-arrange the formula given above to solve for the microstrip line width, you get:

$$w = 1.25 \left[(5.98d)e^{-\left(\frac{Z_0\sqrt{\varepsilon_r+1.41}}{87}\right)} - t \right]$$

where w is the microstrip width (in mils), d is the dielectric thickness (in mils), t is the trace thickness (in mils), ε_r is the dielectric constant, and Z_0 is the desired characteristic impedance. Table 3-4 lists microstrip line widths for the most common applications likely to be used. Note that a few mils of error on your board will not dramatically shift the impedance; a ten percent tolerance on characteristic impedance is usually "close enough."

Table 3-4: Common Microstrip Widths

Board Material	Board Thickness (inch)	50 Ω Microstrip Width (mils)	75 Ω Microstrip Width (mils)
FR-4	1/8	233	107
FR-4	1/16	115	53

There are a few last details to discuss about microstrips. In addition to the approximations discussed above (infinite ground and nothing above the line), there are two design guidelines to keep in mind. The first is to keep other traces, either on the same or other layers, as far from the microstrip as possible. A good rule of thumb is to keep all other traces and vias at least five line widths from the microstrip. This will not always be possible, but it is a good goal. Another design guideline is to try to make all of your RF traces into microstrips. This can be as easy as setting your default routing width to your desired microstrip width. In this way, every RF trace you route (no matter how short) will automatically have the proper impedance.

Note that all of the transmission line calculations and tables given in this manual are approximations. When you have an RF circuit board manufactured professionally, be sure to ask for and follow the recommendations of the fabrication house. They should be able to provide you with recommended microstrip width and dielectric thickness for transmission lines of a given characteristic impedance. These recommendations will take into account all of the details of their fabrication process, and should almost always result in properly designed transmission lines. Note also that you, as the board designer, can request the fabrication house to test the impedance of their boards. These tests are usually performed with a TDR (time-domain reflectometer), an instrument which measures reflections on transmission lines. The usual procedure is that the fabrication house will design a test coupon that is built along with your circuit board. The test coupon will allow the fabrication house to quickly and easily measure the characteristic impedance of the microstrips on your board.

Special RF Routing Techniques: Stripline

Stripline is another type of transmission line that can be easily built on a circuit board. It is identical to microstrip, but with ground planes both above and below the trace. Figure 3-7 shows a cross-sectional diagram of stripline. Stripline offers much improved isolation over microstrip, but at the cost of increased RF loss. Striplines are most often used for either high- or low-level RF signals requiring isolation from surrounding circuitry.

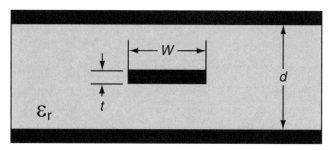

Figure 3-7: Cross-section of a stripline

As with microstrip, several simplifying approximations allow one to write a simple formula for the characteristic impedance of the stripline. This equation is:

$$Z_0 = \frac{60 \ln\left[\dfrac{1.9d}{(0.8w + t)}\right]}{\sqrt{\varepsilon_r}}$$

where w is the width of the stripline trace (in mils, which are 0.001 of an inch), t is the thickness of the trace (in mils), d is the total distance between ground planes (again, in mils), and ε_r is the *dielectric constant* of the dielectric material used in the circuit board.

Once again, this equation can be re-written to solve for the stripline width, given a desired characteristic impedance:

$$w = 1.25\left[(1.9d)e^{-\left(\frac{Z_0\sqrt{\varepsilon_r}}{60}\right)} - t\right]$$

Striplines behave identically to microstrip, but with the added benefit that the RF signal is surrounded top and bottom by ground. The ground planes provide a high degree of isolation, so external signals are less likely to interfere with the RF signal on the stripline. The reverse is also true; RF signals on the stripline will radiate much less energy due to the shielding effect of the ground planes. The downside to stripline is increased RF loss. This is due to the fact that the dielectric (insulating) material is now on both sides of the trace, and tends to absorb more of the RF. Table 3-5 lists stripline widths and decibel loss rates for a few stripline designs.

Table 3-5: RF Losses for Various Striplines

(1/16 inch thick each layer, d=1/8 inch)	50Ω Microstrip Width (mils)	Signal Frequency (MHz)	RF Loss Rate (dB/inch)
FR-4	53	1	0
FR-4	53	1000	0.11
FR-4	53	10000	1.06
Ceramic	21	10000	0.15
Teflon	103	10000	0.05

Table 3-6 lists stripline line widths for the most common applications. As with microstrips, you should discuss your stripline requirement with your fabrication house for the most accurate line width to use for a given characteristic impedance.

Table 3-6: Common Stripline Widths

Board Material	Thickness Between Groundplanes (inch)	50Ω Microstrip Width (mils)	75Ω Microstrip Width (mils)
FR-4	1/4	110	42
FR-4	1/8	53	20
FR-4	1/16	25	9

Striplines are most easily constructed on the inner layers of multi-layer printed circuit boards. If you are making a multi-layer board, then building a stripline layer is easy. First define two ground layers (one on either side of your stripline layer), and then define your stripline layer in the middle. Remember to set the default routing width on the stripline layer to the stripline's design width. You will have to use vias or through-hole device pins to connect to and from the stripline layer.

Note that the stripline layer can also be used to route other (non-stripline) signals. Just remember to keep all signals far from the striplines. In fact, it is best to place copper (ground) areas around your striplines, right on the stripline layer. This will enclose the RF trace in ground on all sides, and will provide the highest degree of isolation available. The resultant stripline trace is very similar to a coaxial cable, with the signal in the center, completely surrounded by ground. Remember when placing the ground areas to keep them at least five line widths away from the stripline. Use plated-through vias to "tie" the ground areas to the ground planes above and below the stripline layer. Figure 3-8 shows an example stripline layer with ground isolation around each stripline. Notice in the figure how the RF striplines are completely protected by ground areas which are kept far from the stripline trace. Also note that there are other ground areas which are around non-RF traces. These ground areas are allowed to come much closer to the traces, because they are not controlled impedance transmission lines. The circles represent vias, in which signals are crossing through this layer from other layers.

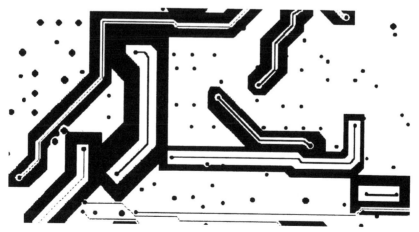

Figure 3-8: Stripline layer with ground isolation areas

Just because you are building your own two-layer circuit boards does not mean that you cannot use striplines. Simply etch the stripline onto one side of a board, with a ground plane on the other side. Then use another single-sided board (of the same thickness) as the ground plane for the other side of the stripline. Place the two boards together, with the stripline traces touching the empty (non-copper) side of the single-sided board. Drill many holes through the boards in the ground areas (don't drill through your striplines!), and then use press-in vias or small wires to solder the two ground planes together. Finally, wrap copper tape around the outer edges of the circuit board and solder the tape to both ground planes. Now you've made your own stripline circuit board!

CHAPTER 4

Real World Guidelines for Commercial Fabrication Houses

This section describes the information that a professional fabrication house needs to know about your design in order to produce boards from your design. This chapter contains all the "real world" information you need to get it done right. You'll also find this information useful even if you are making your own boards by hand.

Design Guidelines

One of the first things a fabrication house will ask you when you contact them about fabricating your board is what the board design parameters are. As a minimum, what they want to know is:

1) The minimum width of the traces you have used.

2) The size of your smallest via.

3) The minimum spacing between the traces you have used.

4) The minimum spacing between the pads you have used.

They may also want to know:

1) The minimum spacing between traces and vias.

2) The minimum spacing between traces and pads.

3) The minimum spacing between vias.

4) The minimum spacing between vias and pads.

The best thing you can do when releasing a board to a fabrication house is to send them the design files. They can extract all the information right from the design files and also check them for any errors. Let's go through all the parameters contained in the design files and see what they mean.

🖐 *Minimum trace width:* Circuit board fabrication houses routinely build boards with trace widths of 4 mils or less. A mil is 1/1000th of an inch. Be warned, however, that the smaller the trace width gets, the lower the yield the fabrication house achieves and the more costly your boards will be. For boards fabricated by a professional fabrication house, I would not use a trace width less than 6 mils unless forced to. You should talk to your fabrication house to find out what their desired design parameters are.

For boards made with the process outlined in this manual, I would use a minimum trace width of 20 mils. Neck traces down to 10 mils to fit between pins if you have to. The wide traces insure that you don't experience "break out" when you etch your board. If your trace width is below 10 mils, it is possible for the etch resist layer to over develop or flake off when you etch, thus breaking the connection and giving you a big headache!

🖐 *Via size:* Unless there is a really good reason not to do it, use a single via for your entire design. Via size is defined as a pad size and a drill size. For example, a pad size of 30 mils and a drill size of 18 mils is commonly referred to as an 18C30 via (the C means circular). Circuit board fabrication houses routinely build boards with via size of 30 mil pad and 18 mil drill. Vias can be much smaller with techniques such as laser drilling but you will pay dearly for them. For boards fabricated by a professional fabrication house, I would not use a via size less than a 30 mil pad and 18 mil drill. You should talk to your fabrication house to find out what their desired design parameters are.

Since plated-through vias are out of the question when "hand fabricating" boards, this book uses *barrels* instead of vias. A barrel is placed through the circuit board and connects a trace on one side of the board to a trace on the other. For the barrels listed in the resource list, use a via pad of 60 mils *minimum* with a drill hole of 35 mils. This will allow you ample room to insert the barrel, swage, and solder it in order to make a good connection. The larger pad allows you to have more leeway in your layer registration, thus increasing your yields. The via pad size is large, relatively speaking, but you can use the barrel as a test point when you debug your design. Eyelets are available from the companies listed in Chapter 7.

    ✥ *Minimum trace spacing:* Circuit board fabrication houses routinely build boards with trace spacing of 4 mils or less. As with trace width, the smaller the spacing, the lower the yield the fabrication house achieves and the more costly your boards will be. For boards fabricated by a professional fabrication house, I would not use a trace spacing less then 6 mils unless forced to. You should talk to your fabrication house to find out what their desired design parameters are.

For boards made with the process outlined in this book, I would use a minimum trace spacing of 10 mils. Spacing less than 10 mils may result in bridging, or shorting between adjacent traces, vias, or pads.

    ✥ *All other minimum spacing:* Professional board houses can handle boards with all spacing 4 mils or less. However, the same warning between spacing and cost applies as described above. For boards fabricated by a professional fabrication house, I would not use any spacing less than 6 mils unless forced to. Again, you should talk to your fabrication house to find out what their desired design parameters are.

For boards made with the process outlined in this book, I would not use a spacing of less than 10 mils. Spacing less than 10 mils may result in bridging, or shorting between adjacent traces, vias, or pads.

Let's now examine some general principles that will make your life easier. These guidelines are geared to the professional circuit board fabrication houses as well as the procedure outlined in this book:

    ✥ For easy-to-read silkscreen on a professionally fabricated circuit board, make the text height 100 mils and the stroke width 10 mils. If you have a circuit board with lots of surface mount components or closely spaced components, you can make the silkscreen size 80 mils with a stroke width of 8 mils. However, this is as small as you want to go. Anything smaller will just end up as an illegible splotch of ink on the circuit board.

✍ Another helpful hint to make your circuit board look more professional is to keep all the text height and stroke widths the same. Unless you have a good reason to do otherwise, try not to have 100 mil characters in one section and 200 mil characters in another.

✍ Make all pads for components as large as possible. This allows you to have more leeway in your layer registration, thus increasing your yields. For increased manufacturability and solderability, the pad size minus the hole radius should be at least 20 mils. This 20-mil ring is referred to as *annular ring*. Professional circuit board manufacturers can produce boards with as little as 3 mils of annular ring. Remember that the smaller the annular ring, the lower the yield and the higher the cost.

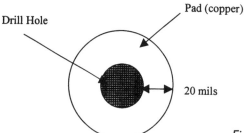

Drill Hole

Pad (copper)

20 mils

Figure 4-1: Recommended pad size

✍ When soldering components to a double-sided circuit board, solder to both sides of the board. This is necessary since we do not have plated-through holes. If you have a component that is hard to solder on both sides, such as a through hole to PLCC socket, try installing and soldering barrels (eyelets) into the board first. This way you will only have to solder the component from the back side.

✥ If you are having your boards professionally fabricated, your silkscreen layer should define the part orientation sufficiently. If you are making your own boards, you won't have a silkscreen layer so you need to make the pad for pin 1 of each component slightly larger or different than the rest. This way you will be able to correctly orient the part on the circuit board without having to refer to the artwork. My preference is to make pin 1 square and the rest round.

Sample Fabrication House Guidelines

In order to give you an idea of what the design rules from a commercial fabrication house can look like, I have included the design rules for two fabrication houses: Nexlogic Technologies and Capital Electronics, Inc (see Chapter 7 for more details). I am providing the guidelines for both so that you can compare the two and see how the guidelines for two fabrication houses differ.

Note that the guidelines are further differentiated into *standard* and high density processing (just to make matters more confusing). The effective difference between the "high density processing" and the "standard processing" to the consumer is strictly cost. If a board can be designed with the rules as described in the standard processing sections, the cost will be lower as compared to the high density processing. Note that there is no differentiation noted between analog, digital, or RF type placements. This is because the fabrication house really doesn't care. You could send them a board layout with nothing on it but your name in copper and as long as it doesn't violate any of their rules, they will manufacture it. It is up to you to make sure that your designs are logically and carefully placed, routed and reviewed.

Remember that a fabricator can update its guidelines listed at any time. You should always check with your design house to make sure you have the most up-to-date information.

Nexlogic Technologies Design Guidelines

Standard Processing

Standard line and space	6 mils
Minimum finished via size in .062" material	8 mils
Minimum unplated hole	8 mils
Minimum finished board thickness	15 mils
Standard overall board thickness tolerance (.062")	± 7 mils
Standard plated through hole size tolerance	± 3 mils
Minimum trace width tolerance	± 1 mil
Standard routing tolerance	±10 mils
Scoring location tolerance	± 10 mils
Standard scoring depth (.062")	20 mils ± 5 mils
Standard warpage tolerance	10 mils/inch
Minimum solder mask lay down	3 mils
Minimum solder mask clearance	1.5 mils

High Density Processing

Minimum line and space	3 mils
Minimum finished via size in .062" material	8 mils
Minimum unplated hole	8 mils
Minimum finished board thickness	15 mils
Minimum overall board thickness tolerance	± 3 mils
Minimum plated through hole size tolerance	± 2 mils
Minimum trace width tolerance	± 1 mil
Minimum routing tolerance	± 5 mils
Scoring location tolerance	± 10 mils
Standard scoring depth (.062")	20 mils ± 5 mils
Minimum warpage tolerance	7 mils/inch
Minimum solder mask lay down	3 mils
Minimum solder mask clearance	1.5 mils

Capital Electronics Design Guidelines

Standard Processing (single- and double-sided boards)

Minimum line width and spacing	10 mils
Single-sided board, pad to hole sizing	+38 mils
Double-sided board, pad to hole sizing	+22 mils
Smallest plated through hole size	24 mils
Smallest non-plated hole	28 mils
Solder mask clearance around pads	+20 mils (screened)
Solder mask clearance around pads	+8 mils (photoimaged)

High Density Processing (double-sided boards)

Minimum line width and spacing	6 mils
Smallest plated through hole size	13 mils
Pad to hole sizing	+15 mils
Solder mask clearance around pads	+12 mils (screened)
Solder mask clearance around pads	+5 mils (photoimaged)

Multi-Layer Processing (standard/high density)

Pad to hole sizing — outer layers	+22/+15 mils
Pad to hole sizing — inner layers	+28/+16 mils
Inner layer plane clearance to hole	+30/+18 mils
Smallest plated through hole size	22/12 mils
Minimum line width — outer layers	10/6 mils
Minimum line width — inner layers	10/8 mils
Minimum line spacing — outer layers	10/6 mils
Minimum line spacing — inner layers	10/8 mils
Solder mask clearance around pads	+22/+12 mils (screened)
Solder mask clearance around pads	+10/+5 mils (photoimaged)

The preceding guidelines can be used for determining the most economic and expedient method of designing a printed circuit board for manufacturing. These guidelines are based on the IPC-D-320A, standard tolerances and manufacturing limitations as listed below.

PRODUCT	STANDARD	HIGH DENSITY
Hole size, for holes under .072" diameter	± 3 mils	± 2 mils
Hole size, for holes over .072" diameter	± 5 mils	± 4 mils
Annular ring	3 mils	1 mils
Circuitry registration	± 5 mils	± 3.5 mils
Registration of inner layers (per linear inch)	± 5 mils	± .35 mils
Trace width reduction	15%	10%

Generation of Design Files

Once you have finished your design, how do you send the design to a fabrication house? If you are going to build you own circuit boards using the procedure outlined in this book, you can skip this section. If you are going to send your circuit boards to a professional fabrication house, you need to generate *design files* that the fabrication house can understand.

Since there are dozens of schematic capture and layout tools on the market, there must be a standard file format that any fabrication house can utilize. This standard file format is known as *Gerber*. Any decent layout tool must be able to produce Gerber files. If yours doesn't, get one that does!

Once you generate Gerber files for your design, it is a good idea to check them with a viewer other than the software that generated them. This helps to catch most of the silly problems. If you don't have a decent Gerber file viewer, you can download a free version of GCPrevue from GraphiCode, Inc. (http:\\www.graphicode.com)

As with any standard, there are always different flavors. Currently there are really only two main flavors: "standard Gerber" and "extended Gerber." RS-274-D is standard Gerber and requires an external aperture list to define all the apertures in the design (more on that in a moment). RS-274-X is extended Gerber and contains all the aperture definitions needed by the design. In other words it is a self-contained file. The preferred flavor is extended Gerber. This really is all you need to know about Gerber files, but if you are a stickler for details read on.

A Photoplotter Tutorial

In order for a circuit board fabrication house to produce your boards, they must expose the *resist layer* of a circuit board with the image, or images, of your design. The resist layer controls the removal of copper during the etching process. (Refer to Chapter 5 for more information on the resist layer.) In this book, we use computer printouts as a means to expose the resist layer. A professional board house will use a photoplotter to create photoplots. You can think of a photoplot as an extremely detailed computer printout on transparent material (called *film*). The film develops just like regular photographic film. The extreme resolution of film is the reason a professional fabrication house can produce circuit boards with such high resolution and small features. Just what is a photoplotter? A photoplotter is simply a plotter that writes with light. A photoplotter uses your design files (Gerbers) to decide what to do. Your design files tell the plotter:

1) What "tool" to use.

2) When to use the "tool" and when not to.

3) Where to go next.

4) Whether to travel in a straight line or an arc.

A "tool" for a photoplotter is a specially shaped aperture, through which light can pass and create a given shape on a piece of film. An aperture is a small window that controls the passing of light similar to the one in a camera.

There are two main types of photoplotters: *vector* and *raster* (bitmap). Very few vector photoplotters are in use any more. Most are raster; however, the vector information makes for a good history lesson.

Vector Photoplotters

Vector photoplotters write with light. You can think of the pen rack on a pen plotter similar in function to an aperture wheel on a photoplotter. The aperture wheel is a disk with 24 or 70 apertures arrayed radially along its circumference. When the photoplotter selects an aperture, the aperture wheel is rotated in order to place the desired aperture between the light source and the film. Apertures themselves are pieces of film and can be made to resemble any shape.

Passing light through an aperture without moving the aperture is known as a *flash*. Moving the aperture is known as a *draw*. However, flash and draw apertures cannot be used interchangeably.

The setup of an aperture wheel is an exacting and time-consuming process. Each aperture on the wheel must be hand mounted and aligned. In order to control setup costs, designers and fabrication houses have agreed upon a standard set of apertures that will be used. Standard Gerber files were not designed to communicate any information about the shape of the aperture in use, only to specify where they are used. This typically leads to a great deal of confusion between the designer and fabricator since designers are not always aware that a separate aperture file must be included with the Gerber file. Worse than that, a designer can easily send an aperture file that is out of sync with the Gerber file, thus creating a bunch of useless circuit boards (not that I have ever done that!).

Raster Photoplotters

The raster, or laser, photoplotter is replacing more and more vector photoplotters. It is really nothing more than a highly precise laser printer. In this situation, the aperture wheel has been replaced with the aperture list. Since these apertures are generated electronically, there is no need for the designer to stick to pre-defined apertures. Some advantages of aperture lists are:

1) Aperture shapes are generated in software, thus providing the designer with greater flexibility.

2) The aperture shapes can be described inside the Gerber file. Enter extended Gerber. These files have the aperture definitions integral to the main design file. There is no way the apertures can be misinterpreted. This is the file format that you should strive to deliver to your fabricator.

3) More apertures can be defined on a list.

In raster photoplots, flash and draw apertures can be used interchangeably.

The formatting of information inside a Gerber file, whether standard or extended, is a complex thing. This book is not the place to discuss that. If you have a desire to delve into the details of Gerbers, you can find a multitude of information on the Internet.

CHAPTER 5

Making Printed Circuit Boards

One of the first steps in making a printed circuit board is to generate an artwork for each signal layer of the board. In this book we will use a maximum of two artworks—one for the top or component side of the circuit board, and one for the bottom or solder side. Artworks used in this book are simple paper plots (1:1 or actual size) of the signal layers. The artworks generated should be positive, meaning there should be black where you desire copper to be on the finished board, and blank where you desire no copper.

There are many ways you can generate your artworks. You can use anything from artwork tape to the Windows® Paint program to simple CAD programs. Anything that you can print or copy can be used as an artwork. I use SuperPCB by Mental Automation to develop my artwork (refer to Chapter 7). You can copy artwork foils from magazines using a photocopier or scanner and use them directly. However, if you use a scanner you can clean up the artwork before printing.

Many factors can affect the quality of your printed circuit board. For instance, the higher the quality of your printer, the finer line resolution you can achieve on your circuit board. I use a Hewlett-Packard LaserJet at 300 dpi and routinely produce boards with line widths of 10 mils. If you use an ink jet printer, you might lose some resolution or sharpness while printing. A dot matrix printer isn't recommended for line widths less than 50 mils.

The thinner the paper you use, the shorter your exposure times will be. However 16-weight paper is about as low as you want to get. If you use lower weight paper, it will tend to stretch and slip as it proceeds through the printer. This will cause layer registration problems on a two-sided board. Registration is nothing more than the alignment of the through-holes from one side of the circuit board to the other. Figure 5-1 contains a section of an artwork created with SuperPCB.

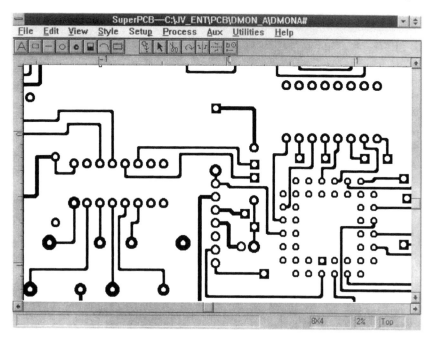

Figure 5-1: SuperPCB sample artwork file

Also, make sure you don't wrinkle the paper as you handle it. Wrinkles and stains will adversely affect the quality of your finished product.

In order to generate the sharpest and finest lines possible when you etch a printed circuit board, you want the toner (black stuff) on the artwork (paper) to be as close as possible to the copper surface of the printed circuit board when the resist layer is exposed. For reasons that will become clear in the next section, this means that we need to print the top layer artwork mirrored and then flip the page over so that the toner rests directly against the copper surface of the printed circuit board. If we were to print the artwork normally, and then expose the resist, we would get some light diffusion due to the thickness of the paper. This would cause the etched traces to be slightly fuzzy and would not produce sharp, clean edges. It would also undercut thin traces possibly breaking them. By mirroring the top layer artwork, the toner rests securely against the copper and does not allow any diffusion of light during the exposure process. If you are producing a two-layer circuit board, the bottom layer can be printed normally. Refer to Figure 5-2 for an example.

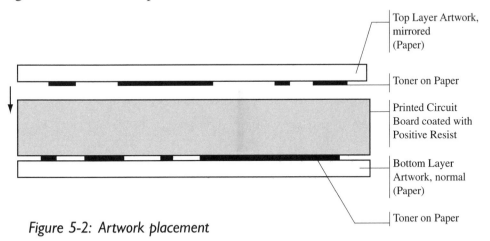

Figure 5-2: Artwork placement

When making a single-sided PCB, your artwork should represent the bottom side (or solder side) of the circuit board. Through-hole components will be inserted from the bare side through to the side with your copper circuitry. When making a single-sided board, your artwork should be printed normally.

Exposing and Developing the Resist Layer

Now that we have artworks, we are ready to expose the resist layer of the printed circuit board.

You may ask "what exactly is a resist layer and what can it do for me?" I am glad you asked! When we etch a printed circuit board, we want to remove copper where we do not want it, and leave copper where we do. In order to do this, we need some way to make the copper "resist" the etchant where we want the copper to stay. If we didn't, when we etch the board, we would end up with a blank, bare board, with no copper at all. This is done with something called a *resist layer*. This resist layer allows us to control the removal of copper during the etching process.

The way the process works is as follows: We start by placing an artwork as close as possible to a copper clad board that is coated with a resist layer. Then we shine a strong, consistent light source at the artwork, resist, and board sandwich. The light penetrates the artwork where it is clear (or in our case slightly opaque), striking the resist and exposing it. This causes a chemical change in the resist layer. The light does not penetrate through the artwork where it is black, or opaque. The resist layer under these dark areas does not chemically alter. When we have exposed the resist sufficiently, we remove the printed circuit board and place it in a solution of resist developer. This developer causes the resist layer that was exposed by the light to dissolve and wash away, thus exposing bare copper. The resist that was not exposed hardens and forms a protective layer that is relatively impenetrable as far as copper etchant solution is concerned. Now when we etch the developed printed circuit board, the bare copper will be removed while the protected copper stays.

How do you get a resist layer onto a copper clad circuit board? I have found the easiest way is to buy the boards "presensitized" with a positive resist. You can just take them out of the bag and use them. The circuit boards I use are made by Kinsten (see Chapter 7), are single or double-sided, are quite inexpensive, and come in a variety of sizes. You can also buy plain copper clad boards and apply the resist layer yourself. It is a simple process where all you have to do is clean the circuit board very well, spray on the resist, and let it dry. You might save a little money, but it will cost you a fortune in frayed nerves and time. Do yourself a favor and buy the presensitized boards.

Whether you sensitize your own or buy ready-made boards, you have to buy the resist developer matched to your particular resist. Since I use Kinsten boards, I use Kinsten developer since it is specially formulated for their boards.

Exposing the Resist Layer

What we need to do is take the artwork and sensitized circuit board sandwich and somehow line up the two layers (if you are making a two-layer board) and hold the whole thing together while it is exposed. This alignment process is usually referred to as *layer registration*. The holding together part is done with something called a contact or exposure frame (those of you familiar with photography might be familiar with the term). For those who are not, a contact frame is nothing more than a hard backboard and a piece of glass clamped together. You sandwich the thing you want to expose between the backboard and glass. This does a great job of keeping everything still and pressed flat. You can purchase a contact frame at your local photography store, or get one specially designed for printed circuit boards (see Chapter 7); however, since I like to build things, I built my own. Plans for our contact frame are included in Chapter 6.

The exposing part can be done as simply as hanging a bulb down over the contact frame. However, if you are going to make more than a few boards you need something better. You need something called an *exposure cone*. Figure 5-3 shows an exposure cone you can build. Plans for our exposure cone are also included in Chapter 6.

Figure 5-3: Exposure cone

The cone helps focus and direct the light and keeps the light source a repeatable distance from the surface of the circuit board. It's important to note here that exposure time varies greatly as the distance from the light source to the board surface varies. This is why it is important to have a reliable and repeatable exposure unit. Another factor in exposure time is the type of light itself. You can pick any type of photoflood light of 150 watts or greater (or any bulb high in ultraviolet). These bulbs are available at photo or hardware stores (the brighter, the better). Once you decide on a bulb, write down its name and where you got it. Otherwise, you will have to recalibrate your system when it comes time to replace the bulb. These bulbs are *very* bright and get *very* hot. Don't stare at it while it is on. Also note that bulb manufacturers recommend that high wattage bulbs be placed with the glass part pointing up, in order to help dissipate the heat. If you hang the bulb from the ceiling, as some others might have you, you may shorten the life of the bulb. If you build the exposure cone listed in the plans section, it will be worth the aggravation. If you do not have the woodworking or electric skills needed to complete the project, you can still get good results by using a store-bought work light from your local hardware store. Buy a work light with a metal reflector around the outside to help focus the light and protect the bulb (and you). Don't forget to make sure that the work light you choose can handle the wattage of the bulb.

If you are still not happy with the exposure methods described above or just don't have the time to build your own system, there is still another option. Various electronic suppliers sell fluorescent exposure systems for about $32.00 (U.S.). These are made to expose the pre-sensitized circuit boards described in this book and their operation couldn't be easier. Exposure times using such systems are about eight minutes.

Now let's get back to aligning both signal layers of a circuit board so that all the vias will line up when the holes are drilled. For those of you making a single-sided board, pay attention anyway, you may need to make a double-sided board some day. The concept of all the vias lining up through the layers of the circuit board is called layer registration. A via is nothing more than a connection from one side of the circuit board to the other side through a hole in the board. Professional board fabricators plate their holes (called plated-through holes) to complete a via. Plated-through holes are too complicated for our labs, so we will use something called a barrel to complete our vias. More on that later.

It is very important that all the pads of the vias line up with their mates so that there is no problem when drilling and inserting the barrels. The way to do this is as follows: Take your bottom layer artwork and place it on a tabletop with the toner facing up. Now take your top layer artwork (which should be mirrored) and place it on top of the bottom layer with the toner facing down. You should now be looking at the backside of the top layer. The toner sides of both artworks should be facing each other. If you now grab the two pieces of paper and hold them up to a window or light, you should be able to see through them. You should be able to line up the pads in both layers precisely! If you can't, check to make sure the top layer is mirrored, and you didn't have any problems printing, such as stretching or distortion, and if you are holding the artworks up to a window make sure it isn't nighttime! If you have built the exposure cone listed in the plans section, you can use the hole in the top of the cone (covered with a piece of glass) as a light table to make the job of alignment easier. When you have aligned all the pads, use tape to attach the two layers together on one side of the artworks only. Now you should be able to move the artworks around without shifting the layers. Take a pair of scissors and trim the artworks to within 0.5-inch of the artwork borders on the three sides that you did not tape. Your artwork assembly should resemble Figure 5-4.

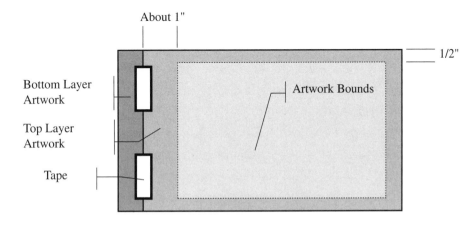

Figure 5-4: Artwork assembly

Now we can insert a pre-sensitized circuit board between the artwork layers and expose it. When you are working with the unexposed circuit board, you need to take a few precautions. Although these boards do react to light, they don't react to low levels of light very quickly. This means that you do not need a dark room to do all this work. A cellar with one small bulb on in the far corner will suffice. We do our work in the daytime and just allow the daylight to enter from three small windows. If you are quick, and do not let sensitized boards hang around you will be fine.

Take a pre-sensitized board out of its bag and *slowly* remove the protective stickers from both sides of a double-sided board, or from one side of a single-layer board. Do not pull too fast or you may remove the resist layer with it! Place the artworks on a flat, clean surface and insert the board between the two artworks with the top layer on top. Insert the board to within 0.5-inch or so from the taped edge. If you are making a single-sided board, just tape the artwork to the top surface. Now tape the top layer to the circuit board in a few places. Make sure that the tape does not overlap any of the artwork toner itself. There is no need to tape the bottom since the contact frame will keep everything pressed and aligned. Leave an exposed section on one side of the circuit board (one where there is no paper). This will be your control section, used to control the resist development process. Now place the board sandwich on the cardboard of the contact frame, place the glass over it and gently tighten the retaining bars (not too tightly or you will crack the glass). Refer to Figure 5-5.

Place the contact frame into the exposure cone so that the surface of the circuit board faces the bulb. Then place the cover on the top of the cone. Refer to Figure 5-6 for details. Now we can expose the topside of the circuit board for five minutes. This time will vary if you use items different from those described in this manual. We use a standard darkroom timer to control the exposure time of our circuit boards. If you don't have one, a watch will do. Varying the exposure time by as much as 30 seconds will not make much of a difference. Now remove the cone's top, take the contact frame back out, remove the glass and the circuit board assembly and flip the board over, being careful not to wrinkle anything. Reassemble and replace the contact frame in the exposure cone and expose the bottom side. The circuit board resist layer is now exposed and ready for development.

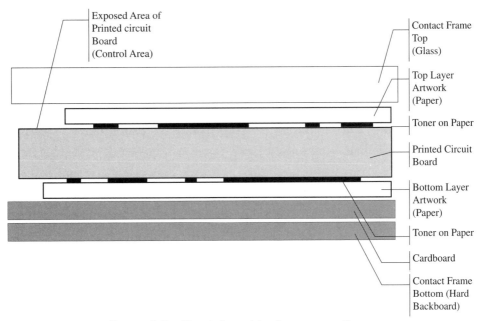

Figure 5-5: Circuit board in the contact frame

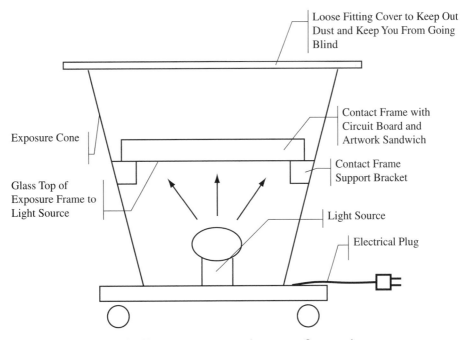

Figure 5-6: Exposure cone and contact frame placement

Developing the Resist Layer

Now we will develop the resist layer. It is important to keep the exposed board away from as much light as possible while developing the resist layer. Take a package of dry resist developer (see Resource List) and empty in a developing tray. Following the manufacturer's directions for the resist developer, add one liter of warm water and mix until dissolved. You must insure that the solution is warm—80° Fahrenheit—not hot, or this process will not work well. That is why the solution is not mixed until needed. Remove the artwork(s) and place the exposed board in the solution and agitate it slightly. You will notice that the exposed resist will "float" off the board as it dissolves, creating a sort of blue "smoke" (if you see real smoke run for it!). Every minute or so, turn the board over to insure both sides develop evenly (and to just see what is happening). Use tongs or wear rubber gloves when you turn the board over. Soon you will see that the positive pattern is starting to show. Continue developing until *all* the resist is gone from the places it should be and you do not see any more blue "smoke" rising from the board. Look at your control section; it will be clear of resist quickly. You should continue the development until all areas that should be clear have the same sheen as your control area. At this point, you can turn the main lights back on.

The manufacturer of the resist developer I use suggests one to two minutes for developing; however, because we used paper instead of film our developing time will be longer. Depending on the temperature of the bath, a complete development could take up to seven minutes. Constantly keep an eye on the board and remove it when complete. If you leave it in the bath too long the developer will start to eat away at the resist that you want to stay. This will also happen if you make the temperature of the bath too hot. If you remove it early, the etchant won't be able to eat away at the exposed copper. Remember that temperature is very important here; I use a water bath to keep the solution at a constant temperature of 80° Fahrenheit. (A rectangular crock-pot makes a good water bath.) You will have to do a little experimenting here to account for all the variances in your procedure. Take notes!

When you are satisfied with the resist development, remove the board and run under water for a minute. This will stop the resist development process. Pat dry with a soft cotton cloth and inspect your work. You can touch up any missed or scratched area with a resist pen (see Resource List), or scrape away any areas you don't want with a razor blade.

Make sure you store your used and unused developer in a plastic or glass container with a plastic cap (not metal). The usable time of solution is one day after use or mixing. Each liter of solution can develop about 20 printed circuit boards (3.9 x 5.9 inches).

You are now ready to etch your circuit board.

Etching the Printed Circuit Board

The process of etching a printed circuit board involves removing the copper that you don't want and keeping the copper that you do. Etching hasn't changed much over the ages. Your only real choice is your etchant chemical.

The traditional choice for etchant has been ferric chloride. Ferric chloride is a very good etchant with a long shelf life. However, it smells and stains everything it touches (including you), except glass. We prefer ammonium persulphate (see Chapter 7) because it mixes easily, is relatively stain free, and doesn't smell. It is also clear so you can see through the solution to see how the etching process is going. The pot life of mixed ammonium persulphate is approximately six months; store in a plastic or glass container with a plastic cap (not metal).

There are two things you can do to an etchant bath to speed up the etching process. You can heat the solution to approximately 100° Fahrenheit, and you can aerate the solution to keep it moving. A plain old fish tank heater works great to heat the solution, and a fish tank aerator with an air stone works great for the aeration.

Before you can etch a board, you need an etchant tank. Do *not* use a metal tank because metals tend to react with etchant chemicals in strange and "interesting" ways. Use glass or preferably plastic. Always wear rubber gloves and use tongs when putting the circuit board in and out of the solution! I use a commercially available etchant tank that I purchased from Circuit Specialists (see the Chapter 7). It costs under $50, contains a heater, aerator, board holder, and a large plastic container with a lid. You can even store the chemicals right in the etchant container. See Figures 5-7 and 5-8.

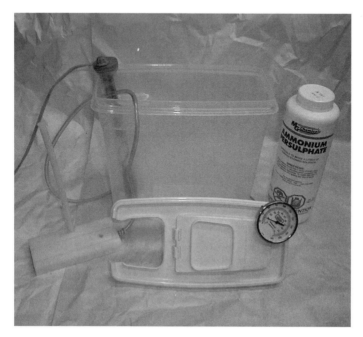

Figure 5-7: Etchant tank and accessories

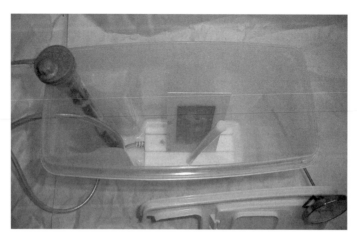

Figure 5-8: Inside view of etchant tank

Now it's time to actually etch a board. Mix a gallon-sized container (clean plastic juice containers work great) of etchant and pour it into your etchant tank. Turn on the heater and aerator and let the bath come up to temperature. When at temperature place the circuit board into your board holder and place the board holder into the tank. At temperature the etching process takes about 15 to 25 minutes. Times will vary depending upon the viability of your etchant solution. Check the board's progress every few minutes. You can tell if the etching process is working if the etchant (ammonium persulphate) turns from clear to light blue. The light blue color is suspended copper. Contact your local town offices or the EPA to find out where you can dispose of the used solution. Used ammonium persulphate is about as reactive as used photography chemicals and can generally be disposed of through a municipal wastewater treatment plant (i.e., down the drain). However, check your local and state environmental regulations before doing so.

Once the etching process has finished, take the board out of the tank with tongs and run it under cool water for a minute. Then pat dry and inspect.

Assuming you are happy with the etched board, the next step is to remove the now unneeded resist layer. You can use a commercial etchant remover (see Chapter Seven), or acetone (local hardware store), mineral spirits, nail polish remover, goof-off (local hardware store), etc. Clean both sides of the circuit board to remove the resist layer. You should now see bright copper. (Don't worry about fingerprints for now.)

We are now ready to tin-plate the circuit board.

Tin-Plating the Printed Circuit Board

Tin-plating is used to protect the copper surface of a circuit board from oxidation and to provide better solderability or solder "wetting." You do not need to tin-plate your circuit board; however, it will provide you with much higher quality and reliability if you do. Professional board houses mostly use electroplating techniques or hot air leveling techniques to plate their board surfaces and through-holes (vias). For high reliability applications, some boards are even coated with gold. However, these techniques are too expensive for most applications, so we will have to try something else.

For example, Datak Corp. produces a product called "Tin Plate" (catchy name). It consists of a chemical bath that you mix, place your circuit board into, and, after some time, it will have deposited approximately 1 to 2 mils of tin-plating (providing you have mixed it correctly and kept it at the right temperature).

Circuit Specialists carries a product called Liquid Tin. This chemical re-quires no mixing or dilution and quickly tin plates copper circuit boards in five minutes or less at room temperature. This is the product I prefer. We will discuss using both products later in this chapter.

In either step, the cleaning process of the board is very important. If the board is not cleaned well, the tin-plating process will result in a splotchy, dull plate. To clean, first remove the resist layer as described in the previous section. Next, take a sponge and sprinkle an abrasive cleanser (such as the "Ajax" brand) on it. Scrub the circuit board for at least 30 seconds on both sides and rinse under running water. Do not use soap or steel wool! The bare copper on the circuit board should be very shiny. If it isn't, keep scrubbing. Try not to let the board dry in the air, as the water evaporates, deposits are left on the board that will affect the tin plate. Also, try not to touch any copper since your finger oils will affect the quality of the plate.

Datak "Tin Plate"

Prepare the solution by mixing 10 ounces of 130° Fahrenheit water with the powder contents of the package. Mix well, then add enough water to make one pint of solution. As with the resist developer and etchant, the tin-plate process is very dependent upon temperature. You want to keep the solution at around 130° Fahrenheit. Lower temperatures will just take longer. At higher temperatures, the solution will not work well.

Place the board in the tin-plate bath using tongs and chemical gloves. Agi-tate the solution slightly and turn the board over every five minutes for double-sided boards. After 20 minutes in the bath, remove the board and clean it again with a cleanser such as Ajax. Put the board back in the bath for another 20 minutes agitation, flipping every five minutes. This should result in a shiny, uniform plating on the surface of the circuit board. If it is splotchy, or not shiny in areas, this is a result of not enough cleaning and there is not much you can do about it at this stage. It will not affect the operation of the circuit board.

Now immerse the tin-plated board in a weak solution of household ammonia and water for 30 seconds (1/4-cup ammonia per gallon of water). This neutralizes the tin-plating, insuring it will not discolor over time.

You can store the tin plate solution for up to six months in a plastic or glass container with a plastic cap. When you reuse the solution, you may notice a white to light yellow precipitate that formed during storage. Most of that "flake" will re-dissolve when you heat and reuse the solution.

"Liquid Tin"

Prepare the bath by pouring the room temperature pre-mixed solution into a plastic tray. Now place the board in the bath using tongs and chemical gloves. Agitate the solution for five minutes. For a double-sided board turn the board over, agitating for another five minutes. Take the board out of the bath with the tongs and run it under cool water for a minute or so. You can pour the used solution back into the original bottle for storage. Now isn't that much easier?

Drilling and Shaping the Printed Circuit Board

Now that we have tin-plated the circuit board, we can drill through-holes, add vias if needed, and shape the board to its final size.

Unless your circuit board is a 100% single-sided surface mount design, you will need to drill holes in it to accommodate your components and mounting holes. There are two basic things you will need to drill your holes

1) A drill bit that is the correct size.

2) Something to hold the drill bit.

Drill bits for printed circuit boards can be as small as 7 mils. You can't drill these holes by hand because I guarantee you will break the drill bit (and I guarantee I will say "I told you so!"). These bits must be held in a drill press. I use a small jeweler's drill press. You could also use a Dremel tool with drill press stand (a Dremel tool is a handheld rotary tool that is inexpensive and does a great job controlling the speed and accuracy of the bits). The bits themselves can be acquired in a multitude of ways. You can get new bits from tool supply houses or used/reconditioned bits at surplus catalog stores. I use mainly reconditioned bits because they have a long life span and are very cheap. I have included many sources for bits in Chapter 7. Don't use regular drill bits because they won't work for long. Circuit boards are made out of a fiberglass resin (FR-4) which is very hard and will dull most normal bits in one or two holes. Bits made for circuit board drilling are made from carbide steel that is very strong and will last for thousands of holes (if you don't snap it off first!).

If you are trying to save money and don't want to buy a lot of different size drill bits, buy a set of the following sizes:

> ♝ For very small devices like transistors (TO-92), use an 18-mil drill.

> ♝ For most everything in the world such as DIP packages, resistors, capacitors, use a 34-mil drill.

> ♝ For large devices such as power transistors, use a 60-mil drill.

> ♝ For mounting holes, use a 125-mil drill.

All you have to do is chuck your bit in the drill, and drill away. Go slowly at first or you will end up breaking the bits. Don't forget to wear your safety glasses. Figure 5-9 shows a jeweler's drill press from Micro-Mark. Also, you should wear a mask while drilling and shaping your boards to avoid breathing the fiberglass dust.

When you are finished drilling, inspect your work. Hole edges should be sharp and clean. If you have rough edges on the holes or the copper pads pulled off of the circuit board, your bit is dull and you should replace it.

Once you have drilled all the holes you need, your next job is to complete all your vias. Vias connect traces from one side of the board through to the other. Professional board houses plate these holes so that there is tin from one side through to the other. While we don't have this technology available to us, we can accomplish the same task a few other ways. If you are cheap (or thrifty, if you prefer), you can do something as simple as inserting a small bare wire—such as

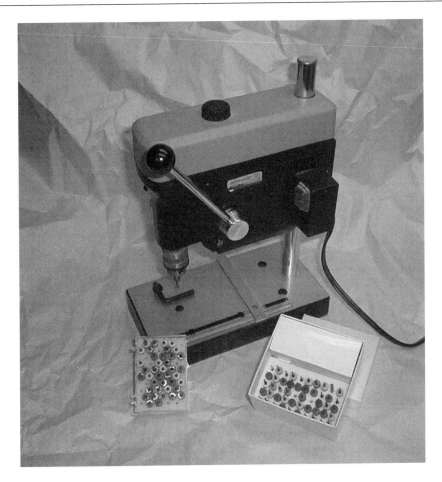

Figure 5-9: Jeweler's drill press with bits

bus wire or a lead from a resistor—into the hole, solder on both sides and trim the wire flush, thus completing the connection. This works well as long as you do not have many vias to complete. If you have many vias, you can use what are called *eyelets*. Eyelets are small barrels that are inserted into a hole from one end and *swaged* (or formed) from the other. You then flow solder around the eyelet to complete the connection. These eyelets cost only a few cents each apiece but require support components such as an anvil and form tool. They come in many sizes and shapes, they do a great job, and look quite professional. This is what we use. Figures 5-10 and 5-11 detail the installation process.

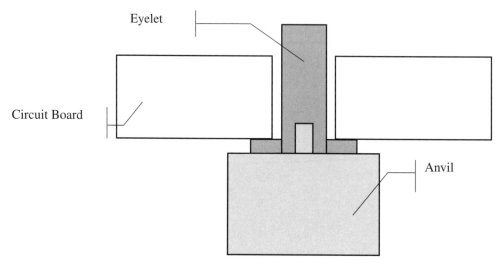

Figure 5-10: Eyelet before swaging

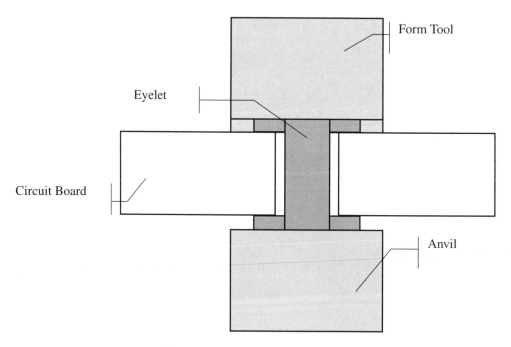

Figure 5-11: Eyelet after swaging

The last step in manufacturing your own circuit board is to shape the board to its final size. Unless you are lucky enough to have the bare board size match your desired finished size, you will have to remove some of the unused FR-4. The easiest way to do this is with a router and a circuit board router bit. I use my Dremel tool with Dremel's router table attachment (see Chapter 7). The router bits themselves were again acquired from a surplus electronics supply house. A router can be very dangerous piece of machinery. Read the directions that come with your equipment. If you don't feel you can perform the task of routing safely, find someone who can. When you are routing, always wear shatter-resistant safety glasses.

Congratulations! You have completed all the steps necessary to make your own high-quality, low-cost printed circuit boards. You may find it takes a few iterations of the steps we've described before you have the process mastered.

CHAPTER 6

Project Plans

Exposure Cone

The exposure cone, mentioned in Chapter 5, is used to hold the exposure frame containing the circuit board/artwork sandwich in order to expose the resist layer. In this chapter, we'll discuss plans for construction of an exposure cone.

Just a few notes before we get to the details:

- The cone shown below includes a light switch to control the bulb. This switch must be able to handle a minimum of 300 watts of AC power. If you are not sure how to wire the bulb and switch—*don't!* Find some-one who can. Remember, electricity can and does kill. If you feel you don't have the skills necessary to build this cone, you can still get good results by going to a hardware store and buying a clamp-on work light that can handle a 300-watt bulb. Just clamp the light fixture to something solid and point towards your exposure frame.

- The bulb socket must be a base capable of handling a minimum of 300 watts of AC power, 600 watts is even better. This means it probably should be ceramic. This base is available at any decent hardware store. The base itself is attached to a piece of sheet metal. This is done because the bulb will get *very* hot, and the metal will dissipate the heat much better than wood. Remember that metal dissipates heat but wood burns!

Building the Exposure Cone

Start by cutting the four sides. The sides are all cut from ¾-inch plywood. The slots in the bottom of the sides allow air to circulate around the bulb and up through the top of the cone. This is very important and will greatly increase the life of your bulb. Refer to Figure 6-1 for the dimensions of the cone's sides.

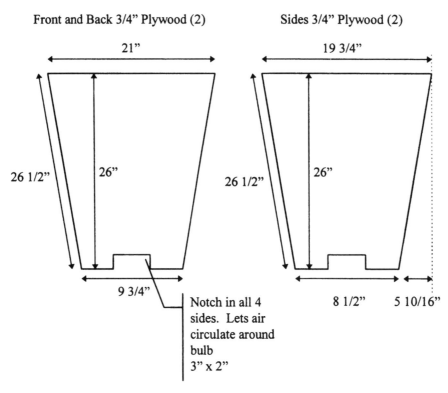

Figure 6-1: Exposure cone side dimensions

Next cut the top piece from ¾-inch plywood. The dimensions are 24 x 24 inches. Cut out the 6 x 6-inch hole in the center. Refer to Figure 6-2 for more details.

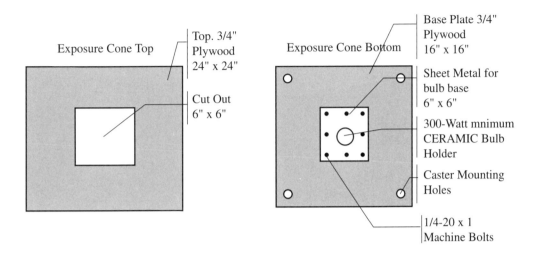

Figure 6-2: Top and bottom plates of the exposure cone

Now cut the bottom piece. The bottom is cut from ¾-inch plywood. The dimensions are 16 x 16 inches. Cut out a 5 x 5-inch hole in the bottom where the sheet metal will go. Also drill the holes for your casters, if you plan on using any. If you do not use casters, you must place blocks under the base to keep the base of the bulb socket from contacting the ground.

Now cut a 6 x 6-inch piece of sheet metal (20 gauge) and drill the hole for the bulb socket. Attach the socket to the metal and attach the metal to the base with 8 ¼-20 x 1-inch machine screws, washers, and nuts.

To assemble the cone itself, start by gluing and screwing the sides together. Pre-drill all the holes so that the wood doesn't split. Try to get the cone as square as possible. Make sure you keep in mind which pieces are the front and back and which pieces are the sides. Next attach the base to the sides. I used right-angle brackets bent slightly to connect the base to the sides; the top just sits on top of the cone. No screws are needed. Refer to Figure 6-3 during this assembly.

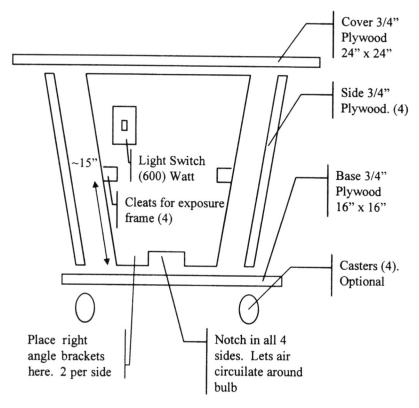

Cover 3/4"
Plywood
24" x 24"

Side 3/4"
Plywood. (4)

~15"

Light Switch
(600) Watt

Base 3/4"
Plywood
16" x 16"

Cleats for exposure
frame (4)

Casters (4).
Optional

Place right
angle brackets
here. 2 per side

Notch in all 4
sides. Lets air
circuilate around
bulb

Figure 6-3: Exploded view of exposure cone

Now attach the casters or wooden blocks, whichever you prefer. If you do not use casters, you must place blocks under the base to keep the base of the bulb socket from making contact with the ground.

Now cut four cleats out of ¾-inch pine approximately 3 inches long and attach to the inside of the cone. These cleats allow the exposure frame to rest securely inside the cone. The cleats should be placed so that the distance from the center of the bulb (filament) to the surface of the contact frame is 12 inches. They should be placed approximately 15 inches up from the bottom of the side. Note that the cleats are slightly larger than they need to be. The larger size insures that the exposure frame will not fall through and smash the bulb. Also note that the frame does not fit tightly into the cone, so that air may flow freely around it. Refer to Figures 6-4 and 6-5 for the cleat details. Note that I glued small stop blocks on the cleats to help center the exposure frame. These are not necessary, but they do help.

Side View　　　　　　　　　End View

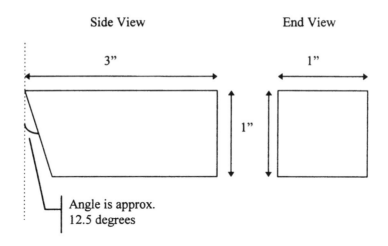

Angle is approx.
12.5 degrees

Figure 6-4: Exposure cone cleat dimensions

Figure 6-5: Exposure cone cleat placement

Wire the light bulb socket to the switch and leave about 4 feet of wire coming out of the switch box. Add a three-terminal type plug to this wire. If you are not qualified to do electrical wiring, call an electrician to do the job!

Paint the whole cone white, except for the bulb socket. This will help to reflect as much light as possible inside the cone, giving you the best exposure, and keep things as cool as possible.

Material List for the Exposure Cone

Here are the materials needed to build the exposure cone:

4	¾-inch plywood sides
1	¾-inch plywood top
1	¾-inch plywood bottom
1	6 x 6-inch sheet metal, 20 gauge
1	600-watt ceramic bulb base
8	right angle brackets, 2 per side
1	electrical switch box
1	electrical switch, 600-watt minimum
1	electrical switch plate
1	three-prong plug
8 ft	12-3 wire
4	casters or wooden blocks
many	1-inch wood screws.
4	¾-inch pine cleats
8	¼-20 x 1-inch machine bolts
8	¼-inch nuts
8	¼-inch washers

Building the Exposure Frame

The exposure frame is used to hold the circuit board and artwork(s) tightly together while exposing the resist layer. The following plans are for a simple frame; refer to Figures 6-6, 6-7, and 6-8 when constructing it.

Figure 6-6: Completed exposure frame

Start by cutting a 12 x 12-inch piece of ½-inch plywood. This will be the base of the exposure frame. You need to cut two slots along the ends of the board. These slots should be 3/8-inch wide x 11 inches long. Locate them 3/8-inch in from the edges. These slots allow the clamping bars to move freely to accommodate different sizes of glass and circuit boards.

Next cut a 10 x 10-inch piece of corrugated cardboard and using double-sided tape or glue attach to one side of the plywood. This will provide a cushion when the circuit board and glass are clamped down.

Now cut the clamping bars. You need 2 ¾ x 1 x 12-inch pieces of pine. Drill 3/8-inch holes in each piece one half inch from the ends. These holes are for the screws.

Finally, paint everything white except for the cardboard. This will help to reflect as much light as possible and keep things as cool as possible.

To assemble the exposure frame, just place the artwork circuit board sandwich on the cardboard, place the glass on top, place the clamping bars on top of the glass, insert, and tighten the screw with wing nuts. Do not over-tighten or you will shatter the glass.

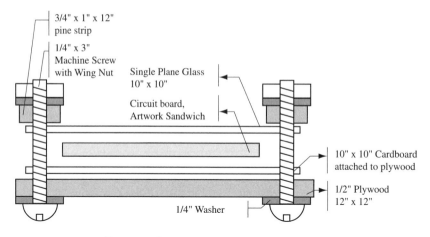

Figure 6-7: End view of exposure frame

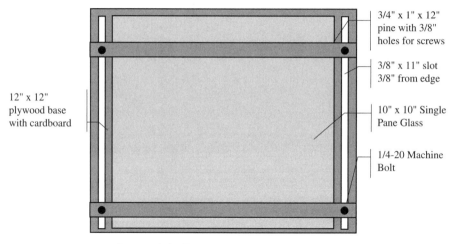

Figure 6-8: Top view of exposure frame

Material List for the Exposure Frame

1	12 x 12 x ½-inch plywood
1	10 x 10-inch corrugated cardboard
1	10 x 10-inch single or double pane glass
2	3/4 x 1 x 12-inch pine strips
4	1/4-20 x 3-inch machine screws
4	¼-inch washers
4	¼-inch wing nuts

CHAPTER 7

Resource List

General Supplies

Contact Printing Exposure Frame. You can either make your own contact frame from the plans in the plan section, or you can purchase a commercial one. A 9 x 12-inch spring-loaded steel printing frame is available from Circuit Specialists, Inc. Current pricing is $14.95. Call or write for a free catalog.

Circuit Specialists, Inc.
PO Box 3047
Scottsdale, AZ 85271-3047
(800) 528-1417
http://www.web-tronics.com

Dremel Tool and Accessories. All of the drilling and routing (shaping) can be done with a Dremel tool. It is available at any major hardware store. If you are not familiar with Dremel tools, it is basically a small hand-held drill with a quick-change collet and a zillion accessories. We use a MultiPro Dremel tool with a drill press stand for drilling and a router table for routing. To find a dealer near you, check out the web page. Major hardware stores carry them.

http://www.dremel.com

Also check out Micro-Mark, The Small Tool Specialists. If you can't find it in their catalog, you don't need it!

Micro-Mark
340 Snyder Avenue
Berkeley Heights, NJ 07922-1595
(800) 225-1066
http://www.micromark.com

Drill Bits. A wide selection of PC drill bits is a must for anyone making their own boards. High-speed steel and carbide bits ranging from 13.5-mils to big are available from many sources. Call or write for a free catalog.

Digi-Key Corp.
701 Brooks Ave. South
Thief River Falls, MN 56701-0677
(800) 344-4539
http://www.digikey.com

All Electronics Corporation
PO Box 567
Van Nuys, CA 91408-0567
(800) 826-5432
http://www.allcorp.com

Electronic Goldmine
PO Box 5048
Scottsdale, AZ 85261
(800) 445-0697
http://www.goldmine-elec.com

Micro-Mark
340 Snyder Avenue
Berkeley Heights, NJ 07922-1595
(800) 225-1066
http://www.micromark.com

A set of 50 reconditioned bits can be purchased for $15.99 from

Northern Tool and Equipment Co
PO Box 1499
Burnsville, MN 55337-0499
(800) 533-5545
http://www.northern-online.com

Etchant, Ammonium Persulphate. Ammonium persulphate is the etchant of choice for JV Enterprises. It has a long pot life (6 months), doesn't stain, and is see-through. Ammonium persulphate is available from Mouser Electronics and Circuit Specialists. Call or write for a catalog.

Circuit Specialists, Inc.
PO Box 3047
Scottsdale, AZ 85271-3047
(800) 528-1417
http://www.web-tronics.com

Mouser Electronics
958 N Main Street
Mansfield, TX 76063
(800) 346 6873
http://www.mouser.com

Etchant, Ferric Chloride. Ferric chloride is the old standby for circuit board etching. It does have more etch capacity than ammonium persulphate. However, it stains everything it comes in contact with. Ferric chloride is available in solution and dry from multiple sources.

Circuit Specialists, Inc.
PO Box 3047
Scottsdale, AZ 85271-3047
(800) 528-1417
http://www.web-tronics.com

JAMECO Electronic Components
1355 Shoreway Road
Belmont, CA 94002-4100
(800) 831-4242
(800) 237-6948 (fax)

Mouser Electronics
958 N Main Street
Mansfield, TX 76063
(800) 346 6873
http://www.mouser.com

Etchant Tank. A handy etchant system that will handle up to two 8 x 9 inch circuit boards at a time. System comes with an air pump for etchant agitation, a thermostatically controlled heater and a tank with lid that hold 1.35 gallons of etchant. Currently priced at $37.95, it is available from Circuit Specialists, Inc. Call or write for a free catalog.

Circuit Specialists, Inc.
PO Box 3047
Scottsdale, AZ 85271-3047
(800) 528-1417
http://www.web-tronics.com

Exposure Kit. This convenient kit includes the major components required to expose presensitized copper-clad boards. This kit contains a fluorescent tube and frame, a clear acrylic weight for the artwork and complete instructions. Currently priced at $31.95. It is available from Circuit Specialists, Inc. Call or write for a free catalog.

Circuit Specialists, Inc.
PO Box 3047
Scottsdale, AZ 85271-3047
(800) 528-1417
http://www.web-tronics.com

Eyelets (Barrels). Eyelets make it much easier to complete a via or plated through-hole. A variety of eyelets, as well as their forming tools are available from International Eyelets, Inc. Call or write for a catalog. When making your own boards, a good eyelet for vias is the F3093-C flat flange eyelet. It costs about $13 per 100. The anvil and form tools are $50 each.

International Eyelets, Inc.
1930 Watson Way, Suite S
Vista, CA 92083
(800) 333-9353
(619) 598-4007
(619) 598-2962

GCPrevue. A quality free-ware program used to view and edit Gerber files. Download from GraphiCode, Inc.

GraphiCode, Inc.
6608 216th Street SW, Suite 100
Mountlake Terrace, WA 98043
(425) 672-1980
(425) 672-2705 fax
info@graphicode.com
http://www.graphicode.com

Gloves. You should always wear rubber gloves and eye protection when dealing with any chemical. You can use dish-washing gloves available at your local supermarket; however, these gloves tend to be thick and bulky. A better choice is a Nitrile disposable glove available from Circuit Specialists, Inc. Current pricing is $ .99 per pair. Call or write for a free catalog.

Circuit Specialists, Inc.
PO Box 3047
Scottsdale, AZ 85271-3047
(800) 528-1417
http://www.web-tronics.com

Ivex WinDraft and WinBoard. Very good, low-cost schematic entry and PCB layout/routing software. The Ivex bundle is currently priced at $349. Runs under Windows. Sold by JDR Microdevices. You can contact them at the address below

> JDR Microdevices
> 1850 South 10th Street
> San Jose, CA 95112-4108
> (408) 494-1400
> (408) 494-1420 (fax)
> http://www.jdr.com

Liquid Tin. Liquid Tin is designed to plate bare copper circuit boards to enhance its solderability. A plate of approximately 1-mil is achieved after only 5 minutes in the bath. Unlike other tin-plating solutions, this one comes premixed, and doesn't need to be heated. It is sold in a 35-oz bottle. Current pricing is $29.50. It is available from Circuit Specialists, Inc. Call or write for a free catalog.

> Circuit Specialists, Inc.
> PO Box 3047
> Scottsdale, AZ 85271-3047
> (800) 528-1417
> http://www.web-tronics.com

Presensitized Positive Circuit Boards. A variety of sizes and types are available from Circuit Specialists, Inc. Prices range from $2.55 for a 3.9" x 5.9" single-sided board to $22.09 for a 12" x 12" double-sided board. Call or write for a free catalog.

> Circuit Specialists, Inc.
> PO Box 3047
> Scottsdale, AZ 85271-3047
> (800) 528-1417
> http://www.web-tronics.com

Positive Circuit Board Resist Developer. Kinsten Industries, DP-50. You will need this developer to develop the resist layer of the positive presensitized circuit boards listed above. It comes in dry form, and you add water. Each 50g package will make 1 liter of developer. Current pricing is $.95 per 50g. It is available from Circuit Specialists. Call or write for a free catalog.

Circuit Specialists, Inc.
PO Box 3047
Scottsdale, AZ 85271-3047
(800) 528-1417
http://www.web-tronics.com

Resist Pen. Resist pens are special markers that resist circuit board etchant. Use them to touch up resist layers. Resist pens are available from multiple sources. Call or write for a catalog.

Circuit Specialists, Inc.
PO Box 3047
Scottsdale, AZ 85271-3047
(800) 528-1417
http://www.web-tronics.com

Digi-Key Corp.
701 Brooks Ave. South
Thief River Falls, MN 56701-0677
(800) 344-4539
http://www.digikey.com

Resist Remover. Commercial resist remover can be purchased if desired. You can also use acetone, or mineral spirits, or nail polish remover. We recommend acetone.

Circuit Specialists, Inc.
PO Box 3047
Scottsdale, AZ 85271-3047
(800) 528-1417
http://www.web-tronics.com

JAMECO Electronic Components
1355 Shoreway Road
Belmont, CA 94002-4100
(800) 831-4242
(800) 237-6948 (fax)
http://www.jameco.com

SuperCAD and SuperPCB. Extremely good, low-cost schematic entry and PCB layout/routing software. SuperCAD is currently priced from $99 to $249. SuperPCB is currently priced from $149 to $249. Runs under Windows. Produced and sold by Mental Automation. You can contact them at the address below.

Mental Automation
5415 136th Place SE
Bellevue WA 98006
(206) 641-2141
(206) 649-0767 fax
http://www.mentala.com

TINNIT Tin-Plate Solution. TINNIT is designed to plate bare copper circuit boards to enhance their solderability. A plate of approximately .5mil is achieved after only 10 minutes in the bath. It is sold in a dry pintsize form. You add water. Current pricing is $4.95. It is available from Circuit Specialists, Inc. Call or write for a free catalog.

Circuit Specialists, Inc.
PO Box 3047
Scottsdale, AZ 85271-3047
(800) 528-1417
http://www.web-tronics.com

Tools, small general. Check out Micro-Mark, The Small Tool Specialists. They carry everything from tweezers to drill bits to calipers to power tools. If you can't find it in their catalog, you don't need it! Call or write for a free catalog.

Micro-Mark
340 Snyder Avenue
Berkeley Heights, NJ 07922-1595
(800) 225-1066
http://www.micromark.com

Commercial Printed Circuit Board Fabrication Houses

(This is a high-quality, low-cost fabrication house)

Nexlogic Technologies, Inc
728 Charcot Ave.
San Jose, CA 95131 USA
Phone: 408-432-8900
Fax: 408-432-8998
Modem: 408-432-8999
http://www.nexlogic.com
sales@nexlogic.com

Quality Fabrication House ($33 prototypes!)

Advanced Circuits
21100 East 3rd Drive
Aurora, CO 80011
Phone: 303-576-6610
Fax: 303-289-1997
http://www.4pcb.com

Design Services

Full design services including design, schematic entry, layout, routing, and fabrication are available from JV Enterprises. Call or write for details.

JV Enterprises
PO Box 370
Hubbardston, MA 01452
Phone: 617-803-3832
jventerprises@att.net
http://www.jventerprises.com

Full design services for RF designs are available from 10-Ring Technologies. Call or write for details.

10-Ring Technologies
PO Box 667
Tyngsboro, MA 01879
Phone: 603-490-9373
TenRingTec@aol.com
http://members.aol.com/TenRingTec

APPENDIX A
Data Monitor Project

Now that you are experts at making your own printed circuit boards, let's apply the information you've learned and build ourselves an extremely useful debugging tool, the data monitor. Figure A-1 shows the assembled unit.

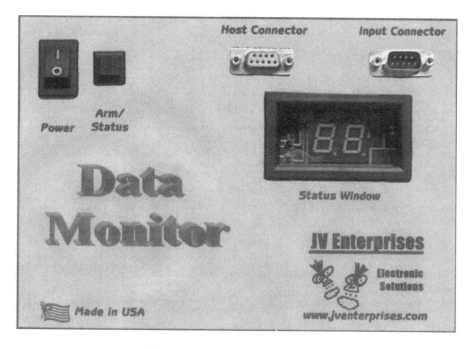

Figure A-1: The Data Monitor

Imagine how useful it would be to be able to monitor and record information about our environment for extended periods of time without having to physically be there. With a variety of input sensors designed for your specific needs, you could monitor and record everything from the temperature of your attic to the number of people who visited your store.

The data monitor described in this appendix is a highly configurable data storage module. It can be used for both short and long term data storage. It supports complex multiple trigger modes. Some of the data monitor's key features are listed below:

✥ Capable of sample and storage times from 30 milliseconds to more than 49 days.

✥ Trigger and storage inputs can include up to four digital and four analog inputs.

✥ EEPROM data storage of 16 kilobits.

✥ Highly error tolerant EEPROM storage; data is retained even if power is removed or lost.

✥ Time and date stamps to insure data integrity.

✥ Multiple trigger modes.

✥ Advanced, eight-term, highly configurable trigger term.

✥ Integrated software.

✥ Low power battery or AC adapter design.

✥ Customizable with different "personality modules."

The data monitor is set up using an easy-to-use Windows program. Once the data monitor has been configured, it runs by itself. When the sample and store process is complete, the same Windows program is used to download all the recorded information.

The data monitor can be configured to monitor and record almost any electronic or electrical circuit through the use of plug-in personality modules. The data monitor described in this appendix is equipped with a general purpose I/O-Temperature Module. This module provides four adjustable gain analog inputs, four adjustable gain digital inputs, and a temperature sensor.

This project contains two double-sided PCBs, perfect for you to apply what you just learned. These appendices contain the artworks for both boards, a schematic, a user's guide, an assembly guide, and application note.

The data monitor exhibits a completely open hardware and software architecture, which is described in full in the subsequent pages. All schematics, firmware and software are provided free from our web page and with the CD included with this book. I also encourage you to write your own software and design your own personality modules. (Note that a complete kit for the data monitor is available from JV Enterprises. See information at the end of this appendix.)

How It Works

The heart of the data monitor is a Microchip Technologies PIC16C74a microcontroller. This microcontroller has some very powerful built-in peripherals such as an 8-channel A/D converter and an asynchronous serial interface. Analog and digital signals are passed to the data monitor from the personality module. The module described in this appendix contains circuitry that buffers the signals and allows you to modify their gains as well as providing a temperature sensor. The temperature sensor signal is applied through one of the analog inputs.

Data is stored and retrieved from the EEPROM using a two-wire technique known as I2C (Inter IC Communications).

Status for the data monitor is shown on a two-digit LED display. The PIC drives these LED segments directly, although the signals are multiplexed. A momentary push button on the front panel enables the display, thus helping to save battery life. The same button also "arms" or begins the sample and store process once the unit is configured.

Serial communications between the data monitor and a PC is accomplished over an RS-232 interface, or COM port. The data and control is transferred at 19,200 bits per second (8 bits, no parity, 1 stop bit). No flow control is utilized.

The unit can run from an AC/DC power "brick" for extended operation or from a 9-volt battery.

System Requirements

The data monitor software will operate on IBM compatible (Pentium microprocessor or equivalent) computers meeting the following requirements:

- Mouse

- 4 megabytes RAM

- Windows 3.1 or higher

- Serial communications port, COM1 (0x03F8, IRQ 4) or COM2 (0x02F8, IRQ 3).

Description of Data Monitor Hardware

This section describes in detail the data monitor hardware as well as the General Purpose I/O-Temperature personality module. The data monitor is shown in Figure A-2. The main features of the data monitor are as follows:

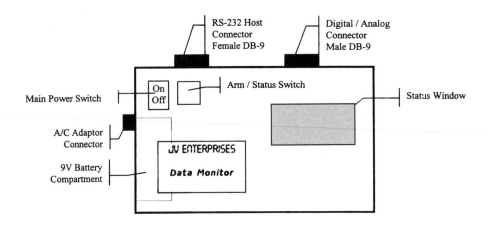

Figure A-2: Data Monitor front panel and features

✎ Main power switch

✎ Status / Arm switch

✎ Status window

✎ A/C adapter connector

✎ Host RS-232 connector

✎ Analog / Digital input connector

✎ 9-volt battery compartment

Let's look at each of these in detail.

Power Switch

The power switch is used to turn the data monitor's power on and off. Turn this switch to the ON position to operate the data monitor. When in the OFF position, the data monitor consumes zero power from the installed battery or AC adapter. When operating from the AC adapter, there should not be a 9-volt battery installed. The converse is also true: when operating from the 9-volt battery, don't connect the AC adapter.

Arm/Status Switch

The Arm/Status push-button switch performs a dual task in the operation of the data monitor. It is also tightly coupled with the data monitor's status window. When the data monitor has first been powered, a press of the Arm/Status switch will cause the status window to display two one-bar characters. This indicates that the unit has passed power-on diagnostics, and is ready to be initialed. With the use of the supplied software, the user can initialize the data monitor to monitor the desired inputs. After initialization, the data monitor waits for the user to arm the unit by pressing the Arm/Status switch. The data monitor will not sample or store any data until the unit is armed. This allows the user to position the unit in the desired location. Once armed, a press of the Arm/Status switch causes the status window to display two three-bar characters to indicate that the data monitor has been armed and is now sampling and/or storing the desired inputs. A press of the Arm/Status switch now will cause the status window to display the current percentage-full of the on-board EEPROM. The percentage will range from 00 to 99 percent. When the EEPROM is full, and the user presses the Arm/Status switch, the status window will display 00. The right-hand decimal point indicates that the EEPROM is full.

Status Window

The status window displays the current status of the data monitor. In order to extend battery life, the status is only displayed when the user presses the Arm/Status switch. When the data monitor has first been powered, a press of the Arm/Status switch will cause the status window to display two one-bar characters. This indicates that the unit has passed power-on diagnostics, and is ready to be initialed. With the use of the supplied software, the user can initialize the data monitor to monitor the desired inputs. After initialization, the data monitor waits for the user to arm the unit by pressing the Arm/Status switch. The data monitor will not sample or store any data until the unit is armed. This allows the user to position the unit in the desired location. Once armed, a press of the Arm/Status switch causes the status window to display two-three bar characters to indicate that the data monitor has been armed and is now sampling and/or storing the desired inputs. A press of the Arm/Status switch now will cause the status window to display the current percentage-full of the on-board EEPROM. The percentage will range from 00 to 99 percent. When the EEPROM is full, and the user presses the Arm/Status switch, the status window will display 00. The right-hand decimal point indicates that the EEPROM is full.

AC Adapter Connector

Power to the data monitor can be supplied in one of two ways: either through the AC adapter, or a 9-volt battery. When using the AC adapter, insure that there is not a 9-volt battery also installed. The AC adapter shipped with the data monitor supplies 9 volts DC at 300 milliamperes. Any replacement adapter must conform to these specifications.

The pin-out for the AC adapter is shown in Figure A-3. The view below is looking into the chassis (mate connector) from the connector (front) side.

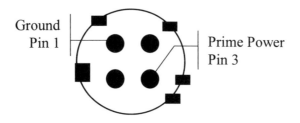

Ground Pin 1

Prime Power Pin 3

Figure A-3: AC adapter pin-out

RS-232 Host Connector

The host or PC communicates to the data monitor through this 9-pin D female connector. It is a simple two-wire interface utilizing the communications port signals RX and TX (pins 2 and 3 on the 9-pin D) and no flow control. This connector should be connected with the supplied cable to the hosts COM1 or COM2 port. The specifications of this link are as follows:

- 19,200 baud
- 8 data bits
- no parity
- 1 stop bit
- no flow control

The data monitor contains a 32-byte FIFO used to buffer the incoming RS-232 messages. The host may send the RS-232 message as fast as possible without any ill effects. However, the host must wait for the response from the current message before beginning another. Because of this feature, there is no software or hardware flow control implemented in this link.

The pin-out of this connector is shown in Figure A-4.

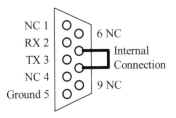

Figure A-4: Host RS-232 connections

Analog/Digital Input Connector

The Analog/Digital input connector connects the data monitor with the outside world. The connections are supplied through the 9-pin D male connector. It allows the four digital inputs and four analog inputs to be connected to the installed personality module. The personality module filters and buffers all eight inputs. Although not necessary, it is recommended that any custom personality modules buffer all the analog and digital inputs. This insures that any "bad things" that may happen do not damage the data monitor in any way. The pin out for the connector is shown in Figure A-5.

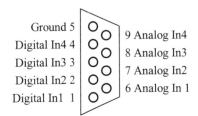

Figure A-5: Analog/Digital connections

The electrical specifications of the analog inputs are given in Table A-1.

Table A-1: Analog Input Electrical Specifications

Parameter	Condition	Min	Typ	Max
Input Voltage Low	Rload <= 10k ohm	0v DC		
Input Voltage High	Rload <= 10k ohm			+5.0v DC
Input Impedance	Rload = ∞ ohm		20k ohm	

The electrical specifications of the digital inputs are given in Table A-2:

Table A-2: Digital Input Electrical Specifications

Parameter	Condition	Min	Typ	Max
Positive Going Voltage	Rload = 5k Ohm	+3.0v DC		
Negative Going Voltage	Rload = 5k Ohm			+2.0v DC
Input Current	Rload = ∞,0 Ohm		+- 1ua	

9-Volt Battery Compartment

As noted before, power to the data monitor can be supplied in one of two ways: through the AC adapter or a 9-volt battery. The 9-volt battery can be any standard Ni-Cad or alkaline battery. When using the 9-volt battery, be sure that the AC adapter is not connected.

Electrical Interface

Electrical connections to a personality module are supplied through the 8- and 10-pin connectors defined in Table A-3 and Table A-4.

Table A-3: 8-Pin Connector Definition

Pin	Name	Description	Direction	Notes
1	VCC	+5 volts supply	Output	20 ma Max
2	VCC	+5 volts supply	Output	
3	V Prime+	+9 - 12 volts supply	Output	100 ma Max
4	Ground	-	Input	-
5	Ground	-	Input	-
6	Ground	-	Input	-
7	Digital 4	Digital Input	Input	TTL +-15ma
8	Digital 3	Digital Input	Input	TTL +- 15ma

Table A-4: 10-Pin Connector Definition

Pin	Name	Description	Direction	Notes
1	Digital 1	Digital Input	Input	TTL +-15ma
2	Digital 2	Digital Input	Input	TTL +-15ma
3	ON/OFF*	Power Control	Output	On(1) Off(0)
4	ID2	Daughter Card ID	Input	Not Used
5	ID1	Daughter Card ID	Input	Not Used
6	ID0	Daughter Card ID	Input	Not Used
7	Analog 4	Analog Input	Input	<10k in impedance
8	Analog 3	Analog Input	Input	<10k in impedance
9	Analog 2	Analog Input	Input	<10k in impedance
10	Analog 1	Analog Input	Input	<10k in impedance

Note that the ON/OFF* pin and the ID(2:0) pins are not used by the data monitor and the ON/OFF* pin is fixed to ON (logic 1).

Personality Module

The data monitor is shipped from the factory with the General Purpose IO - Temperature personality module. This module buffers all the analog and digital inputs to the data monitor, as well as providing a temperature sensor. By turning the data monitor over, and removing the six chassis screws, you can gain access to the personality module. Let's examine the details.

Physical Interface

A personality module is connected to the data monitor's motherboard through 10-pin and 8-pin connectors. For those who wish to create their own personality module, the physical layout of the connectors is shown in Figure A-6. The view is from the backside of the data monitor with the cover and personality module removed.

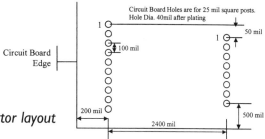

Figure A-6: Personality module connector layout

Analog Interface Circuitry

All analog inputs are buffered though a general purpose op amp circuit. This circuit performs two major functions: it protects the data monitor from unexpected events and can be configured for various gains. It is set to a gain of 1 when shipped, as explained below.

An input voltage of zero volts yields a data monitor A/D result of 0 out of 255. An input voltage of 4.7 volts yields a data monitor A/D result of 240 of 255. Any analog input voltage greater than 4.7 volts will yield a result of 240. If you wish to sample voltages greater than 4.7 volts, change the gain of the op amp circuit to attenuate your input signal, and scale the data monitor results accordingly. The analog input buffers provide an input impedance of 20 kilohms to the host circuit. The diagram for the buffer circuit is shown in Figure A-7.

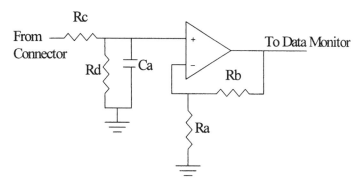

Figure A-7: Analog input buffer circuit

Resistors R_c and R_d form a voltage divider. The module is shipped with these resistors at 10k. This provides a gain of 1/2 or $(R_d / (R_d + R_c))$.

The op amp is configured as a non-inverting voltage follower. Resistors R_a and R_b set the gain of the op amp. The module is shipped with these resistors also set to 10 kilohms. This provides a gain of 2 or $((R_a + R_b) / R_a)$.

Overall, a gain of 1/2 followed by a gain of 2 yields an effective gain of 1. By adjusting these resistor values as needed, the user can insure the maximum range or precision for an analog input.

Capacitor C_a in conjunction with R_c provides a low-pass filter with a cutoff frequency of $1/((R_c \| R_d) C_a)$ or 2000 Hz. The module is shipped with a 0.1 µf capacitor installed.

Digital Interface Circuitry

All digital inputs are buffered through a Schmidt trigger inverter (74HC14). This inverter buffers the input and protects the data monitor from the outside world. The Schmidt trigger provides some hysteresis in the input and insures that the signal does not jitter while transitioning from one logic state to the other. Refer to Table A-2: Digital Input Electrical Specifications, for details. The digital input buffers provide an input impedance of approximately 11 kilohms to the host circuit.

Each digital input also contains a resistor, R_b, which will insure that the digital input does not float when no input is present. All unused digital inputs will be pulled to ground. The pair of resistors R_a and R_b also forms a voltage attenuator, which can be adjusted to attenuate larger signals as desired. The personality module ships from the factory with the attenuation set to approximately 90%, $(R_b / (R_a + R_b))$. The circuit for the digital buffers can be seen in Figure A-8.

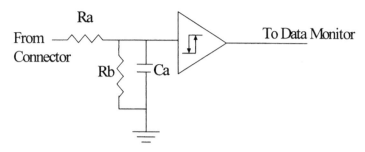

Figure A-8: Digital input buffer circuit

Temperature Module

The General Purpose IO - Temperature personality module contains circuitry to monitor the ambient temperature. This signal replaces the Analog Input 1 signal with the use of jumper J4. Placing J4 on pins 2-3 selects the temperature sensor. Placing J4 on pins 1-2 selects the analog input 1. Figure A-9 indicates the jumper position; this figure is shown from the front (component side) of the personality module.

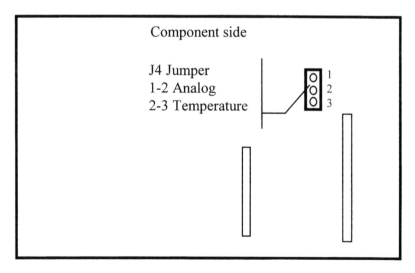

Figure A-9: Personality module jumper locations

The temperature sensor installed on the personality module is a National Semiconductor LM50. This sensor is capable of detecting temperatures from -10 to +65 degrees Celsius. This temperature sensor exhibits a linear output over the full range of temperatures. The output of the LM50 is represented by the following formula:

$$Output\ Voltage = 0.5volts + (10mv)*(Degrees\ Celsius)$$

For example, a temperature of 20 degrees Celsius would yield an output voltage of .5 + (.01*20) or .7 volts.

The output of the temperature sensor is then passed through a non-inverting op amp with a fixed gain of 2.5. This expands the output of the sensor to cover the full range of the data monitor's A/D converter. Remember that the reference voltage for the A/D converter inside the data monitor is 5.0 volts. That means that the digital value reported by the data monitor is represented by the following formula:

$$Output\ value = (Input\ voltage\ /\ 5.0v) * 255$$

Continuing our example, a temperature of 20 degrees Celsius produces an output of .7 * 2.5 = 1.75 volts. This would be reported by the data monitor as an A/D value of 89 out of 255.

Installation and Description of Data Monitor Software

The data monitor software, Dmon.exe, allows the user to set up, initialize, and download from the data monitor. Dmon also allows the user to save and recall custom configurations. All source code for the Dmon application is supplied on the CD accompanying this book.

Description of Data Monitor File Types

Dmon creates and maintains two different data file types during normal execution. The file types are the *configuration files* and the *download files*. Each file type is described in full below.

Data Monitor Configuration Files, .CFG

Dmon allows the user to save and recall configurations for quick and easy data monitor setups. The .CFG files are ASCII text; however, the data in them is encoded. It is recommended that files are not hand edited.

Data Monitor Download Files, .DNL

After the data monitor has monitored the desired information, Dmon allows the user to download the recorded information for processing. This data is saved in a file with the extension .DNL. A sample download file is shown below.

```
Data Monitor Download Data (V1.0) — JV Enterprises

Storage Mode  :: Trigger
Trigger Mode  :: Calculated
Average Off   :: —
Time Stamp    :: 20:08:13.190
Date Stamp    :: 02/09/96
User Time Scale :: 30ms

Day  Time              An1
00   20:08:16.077      93
00   23:09:09.616      91
00   03:56:02.152      89
00   09:41:56.924      91
00   11:12:53.437      93
00   15:00:45.378      95
00   18:26:00.984      97
00   18:26:05.083      99
00   18:26:05.989      97
```

Data Monitor Operation

This section describes in detail the operation of the data monitor through the use of the Dmon.exe program. Before you begin, you should install the software as described in the previous section.

Software Start-up

To start the data monitor configuration software, double-click on the Dmon icon. The Dmon program will then load and start executing. Once the main window is displayed, Dmon is ready for user input. Figure A-10 shows the initial arrangement of the screen.

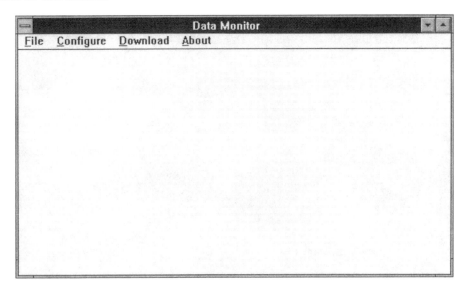

Figure A-10: Dmon initial start-up screen

Data Monitor Dmon Main Window

The only active part of the Dmon main window is the Menu Bar Region. Through these menu pull-downs, the user can open, save, configure, initialize and download the data monitor. All the menu pull-downs are described in detail below.

File PullDown

Open Config

This operation is used to open existing configuration files that the user may have previously created. Select this entry and navigate to the desired directory. By default, only configuration files will show in the selection window. Configuration files end in .CFG. Click OK to load the configuration file, or click cancel to abort the operation. When OK is selected, Dmon will update all internal parameters with the information saved in the .CFG file. The user can check the restored setup by the `Configure->Setup` pop-up.

Save Config

This operation is used to save a current configuration to a user supplied file name. Select this entry and navigate to the desired directory. The user can then select an existing file to replace, or type a new file name in the entry window. Be sure all entered file names end in .CFG. The user will be prompted if an existing file will be overwritten. Click OK to save the configuration file, or click cancel to abort the operation.

Exit

This operation is used to exit the configuration software. Select this entry only when you want to exit the Dmon application. Any unsaved information will be lost.

Configure PullDown

Comport

This operation is used to select the active Comport used to communicate with the data monitor. Select this entry and then choose a Comport. Currently, Dmon supports hardware COM1 (0x03F8, IRQ 4),COM2 (0x02F8, IRQ 3), COM3 (0x3E8, IRQ4), or COM4 (0x2E8, IRQ3).

Setup

This operation is used to set up the data monitor prior to initialization. The initial set-up screen is shown in Figure A-11.

Each major section of the set-up screen is described below.

Storage Inputs

Use these check boxes to select which of the eight Analog/Digital inputs are to be stored during operation. Any or all of the inputs may be selected; however, keep these points in mind:

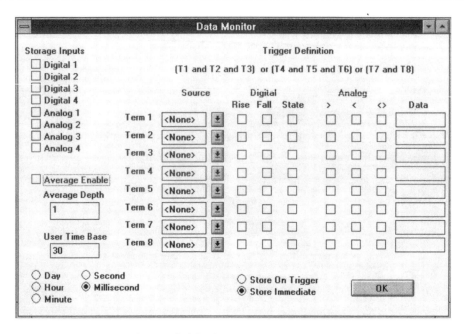

Figure A-11: Dmon setup screen

1. For each analog input chosen, one byte of the EEPROM is used per sample period (this does not include the time stamp bytes, if active). For example, selecting one analog input consumes one byte of EEPROM per sample written. Selecting four analog inputs consumes four bytes of EEPROM per sample written, thus reducing the number of samples per complete cycle.

2. Selecting from one to four digital inputs consumes only one byte of EEPROM space per sample written. Therefore, there is no penalty for selecting more than one digital input as there is for the analog inputs.

Average Depth and Average Enable

Check the average enable check box to enable data averaging. Data averaging is used to average or "smooth" the analog inputs used for storage or for triggers.

The data monitor can be tasked to average all or none of the active analog inputs. The averaging applies to all analog inputs; it cannot be enabled for some and not for others. Once enabled, the data monitor will average all selected analog inputs and use this averaged value for storage and for the trigger modes. The data monitor can average from one to eight (current plus last seven) samples. Enter the average depth desired in the depth box.

Keep in mind a few tidbits about data averaging. After the initial arm of the data monitor, all of the internal average registers are set to 0. Depending on the average depth selected, it will take a few samples for the selected input or trigger to stabilize. Also, the values recorded will be slightly higher or lower than if they were recorded without averaging. For example, consider Table A-5:

Table A-5: Averaging Example

Analog input A/D value	With Averaging of 2	No averaging
Start	0	0
130	0,130 = 65	130
128	65,128 = 96	128
133	96,133 = 114	133
130	114,130 = 122	130
128	122,128 = 125	128
128	125,128 = 126	128
128	126,128 = 127	128
128	127,128 = 127	128
128	127,128 = 127	128
128	127,128 = 127	128

It can be seen that the no averaging mode shown in the "no averaging" column represents exactly the analog input, with all the fluctuations associated with that input. The middle column, headed "with averaging of 2," on the other hand has "smoothed" the analog input, effectively filtering out all the transients. However, this has also affected the result that the data monitor has reported.

User Time Base

This input box allows the user to specify the sample time between data monitor samples. This value is used along with the Store Immediate/Store on Trigger mode defined below. Enter the desired value and choose the appropriate time scale (Day, Hour, Minute, Second, or Millisecond).

Remember to always enter an integer in this field. For example, enter 6 seconds, not .1 minutes.

Store Immediate / Store on Trigger

These mutually exclusive radio buttons determine the storage mode of the data monitor.

Selecting "Store immediate" will use a trigger condition to begin data storage only if one is defined. Otherwise the data monitor will sample and store data based solely on the user's time base. This mode makes the most of the data monitor's EEPROM since it doesn't have to store the time between stores; it is a fixed constant.

"Store on trigger" will sample the inputs as fast as possible (~25ms) and store only those which satisfy the trigger condition. When the trigger is satisfied, the data monitor will store the active inputs along with the elapsed time since the last successful trigger condition. By selective use of this mode bit and the data monitor's trigger, various storage modes can be achieved.

The various trigger modes are defined below:

- ✍ *Store Immediate and no triggers defined*: The data monitor will store the active inputs based on the users time base.

- ✍ *Store Immediate and Triggers defined:* The data monitor will monitor all inputs until the trigger mode is satisfied. It will then store the trigger condition (and associated elapsed time). The data monitor will then fill the EEPROM based on the users time base (no trigger).

- ✍ *Store on Trigger and triggers defined:* The data monitor will monitor all inputs until the trigger mode is satisfied. It will then store the trigger condition (and associated elapsed time). The data monitor will then continue monitoring the inputs and storing data only when the trigger condition is met.

✍ *Store on Trigger and no triggers defined:* This is a slightly illegal mode; however, the data monitor will fill the EEPROM as follows: Because the data monitor has been tasked to store data on trigger but no triggers have been defined, the data monitor will default to storing the active inputs based solely on the user's time base.

Initialize

This operation is used to send the current setup choices to the data monitor. After the user selects this entry, Dmon sends the initialization message to the data monitor. The data monitor then responds with an initialization successful message. At this point Dmon informs the user that the initialization was complete. The data monitor is now armed, and ready to begin monitoring the selected configuration. At this point, the Comport cable can be disconnected, the unit can be placed in the appropriate location, and the Analog/Digital connector plugged in (if needed).

When the user is satisfied with the external connections, and the data monitor is in place, a simple push of the Status/Arm button begins the sample/store mode.

The user can download the data monitor to recover the stored data at any time simply by connecting the data monitor back to the Comport cable and issuing the Download command.

The user should be aware, however, that issuing a Download command interrupts the normal sample/store mode, thereby affecting the timestamps on all subsequent sample/stores. The interruption is on the order of 1 second. Also note that when power is removed from the data monitor, all configuration information is lost. The unit must be re-initialized and armed for storage to begin.

Trigger Definition

The trigger definition controls allow the user to generate a simple or extremely complex trigger condition for the data monitor's sample and store capabilities. The format of the data monitor's trigger is as follows:

(Trigger_1 & Trigger_2 & Trigger_ 3) # (Trigger_4 & Trigger_5 & Trigger_6) # (Trigger_7 & Trigger_8)

The "&" represents a logical AND and the "#" represents a logical OR.

Each of the eight trigger terms may choose any one of the four analog or four digital inputs, and one or more of the trigger modes specified below.

Digital Inputs

    ✎ Digital Input Rising

    ✎ Digital Input Falling

    ✎ Digital State (Level)

Analog Inputs

    ✎ Analog Input >= Level

    ✎ Analog Input <= Level

    ✎ Analog Input Change >= or <= Threshold

For any given trigger, the user may specify one, two, or all of the trigger modes listed above. For example, to specify a trigger of toggle for a digital input you would specify the rising and falling trigger modes.

For the trigger modes of Analog >=, Analog <=, and Analog Change, another word is needed to specify the thresholds or levels. This is specified in the trigger data word. If a trigger does not need this data word, it should be left cleared (empty). Place the decimal value into the corresponding trigger data word.

For the trigger mode of Digital State (level), the corresponding trigger data word should be set to 1 for a logic state of 1 (TTL) and 0 for a logic state of 0 (TTL).

Unused trigger terms should be left blank. Just uncheck any checked radio boxes, and select <NONE> from the trigger term source pull-down menu.

Since the trigger definition can be confusing at first, let's consider an example.

We wish to set up the data monitor so that we can store the ambient temperature whenever any of the following conditions are met:

> 🖑 Digital input 1 rises or Digital input 2 toggles while Digital input 3 is a logic 1.

In our example, we have used trigger terms T1, T4 and T5. Note that all other trigger terms are left unused. An unused trigger term does not factor into the trigger equation. These terms are ignored. For example, if we leave Trigger_2 and Trigger_3 unused, the trigger equation becomes:

(Trigger_1) # (Trigger_4 & Trigger_5 & Trigger_6) # (Trigger_7 & Trigger_8)

The Setup screen should look as shown in Figure A-12:

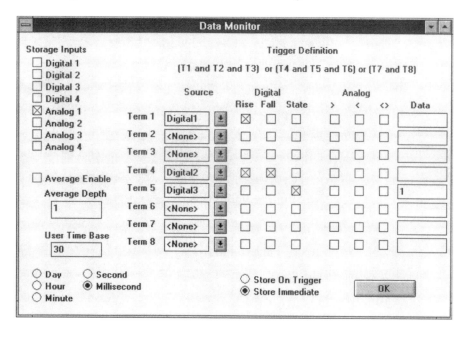

Figure A-12: Example set-up screen

Auto Download Pull-down

This operation is used to configure the data monitor for automatic downloads. The automatic download dialog box is shown in Figure A-13.

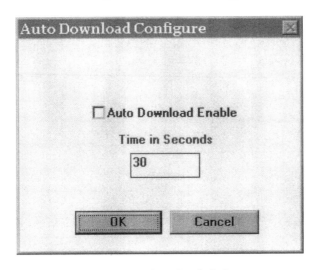

Figure A-13: Auto download dialog screen

When the Auto Download enable box is checked, the data monitor will transmit the current stored EEPROM samples over the serial connection at the specified rate. When the stored data is transmitted, the EEPROM is reset, and the data monitor begins sampling data with the previously specified initialization parameters.

Specify the transmission period in seconds. The minimum value between transmissions is 5 seconds. The maximum value is 4294967 seconds (or ~ 49 days). (Note that in this mode the time and data stamps will be valid for the first download only. Subsequent downloads will not contain the correct time stamp.)

The recommended sequence of events is to set up the auto download parameters, select the stored inputs and trigger conditions, then initialize and arm the unit.

Download PullDown

Download

This operation is used to download recorded information from the data monitor to the host for analysis.

Downloaded information is recorded in a .DNL file specified by the user. It is a ASCII text file and can be pulled into any standard editor or Word Processor. It is recommended that the user import the download file into a spreadsheet such as Microsoft Excel® and use its superior graphics capabilities to plot or analyze the data monitor's results.

Download files contain all pertinent information about the data recorded. A sample download file is shown below:

```
Data Monitor Download Data (V1.0) — JV Enterprises


Storage Mode   :: Trigger
Trigger Mode   :: Calculated
Average Off    :: —
Time Stamp     :: 20:08:13.190
Date Stamp     :: 02/09/96
User Time Scale :: 30ms

Day   Time            An1
00    20:08:16.077    93
00    23:09:09.616    91
01    03:56:02.152    89
01    09:41:56.924    91
01    11:12:53.437    93
01    15:00:45.378    95
01    18:26:00.984    97
01    18:26:05.083    99
01    18:26:05.989    97
```

This sample file tells us that the data monitor was set to store on trigger mode, with a trigger defined (calculated), averaging was disabled, the user time base was 30 milliseconds, and the initialize time was around 8:08:13 PM. Note that the unit was not armed (by pressing the status/arm switch) until 8:08:16 PM. Also note the Day column. This field counts the days elapsed since the store cycle began. Every time the elapsed time passes midnight (23:59:59), the day counter is incremented.

Auto Download

This operation is used to capture automatic download data from the data monitor. This operation should only be used when the data monitor has been configured for automatic download.

Select `Download->Auto Download` to capture a download message that is automatically being transmitted from the data monitor. When selected, the Dmon.exe application will monitor the serial port for 60 seconds looking for an incoming download message. If a message is found, it will be written to the hard disk under the file name specified by the user. If no message is received in the 60-second period, Dmon.exe will abort the search, and inform the user.

When Dmon.exe is searching for an incoming message block, the PC will be locked from the user. Full control will return after the 60-second interval. Once the 60-second interval has begun, there is no way to stop it.

Voltage

This operation is used to monitor the data monitor's main supply voltage by selecting the `Download->Voltage` menu item.

When operating from the 9-volt battery, this voltage can be used to predict the battery life. A fully charged 9-volt battery will return a voltage of around 8.8 to 9.0 volts. A discharged battery will return a voltage of around 7.0 to 7.2 volts.

When operating from the AC adapter, the data monitor will return a voltage of around 11.0 to 12.0 volts. Note that even though the AC adapter is rated for 9-volts DC output, the data monitor will report 11.0 to 12.0 volts. This is because the AC adapter does not have a regulated output. It is merely a transformer. The AC adapter's output voltage is rated under full current load, and the data monitor draws very little DC power.

Current Values

This operation is used to instruct the data monitor to sample all its analog and digital inputs, and display the current values. Select the `Download->Current Values` menu item to activate the Current Values dialog box. The Current Values dialog box is shown in Figure A-14.

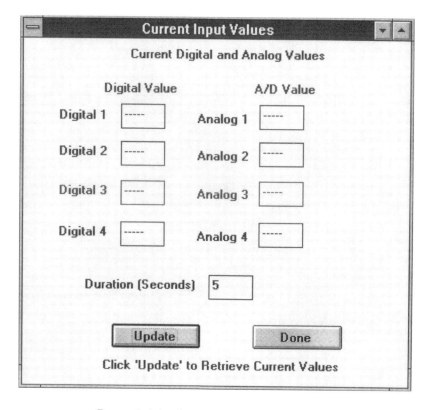

Figure A-14: Current values dialog screen

The analog and digital values are initially displayed as "—". When the Update button is clicked, the data monitor samples the analog and digital inputs, and the values are displayed in the dialog window. The values are updated until the time specified in the Durations window has expired. The user can enter any duration from 0 to 60 seconds. Click the Done button to close the Current Values dialog.

Remote Arm

This operation is used to begin the sample and store cycle of the data monitor. This software command has the same effect as a press of the Arm/Status switch.

To perform a remote arm, select the Download->Remote Arm menu item. The user can arm the data monitor by either pressing the Arm/Status switch, or by using this software command. However, in both cases the data monitor must have been previously initialized or the Arm command will fail.

About PullDown

About

This operation is used to view the current revision level of the data monitor Dmon software. Select this entry to view this information. When finished, close the window with the window's upper left button.

Message Structure

Although somewhat technical in nature, it is recommended that the user read the following sections in order to achieve a greater understanding of the data monitor's operation.

This section is for those who wish to write their own software to communicate with the data monitor. It defines how communications are handled to and from the data monitor, their exact format, and any and all details you probably will ever need to know.

Communications to and from the data monitor are message based. Each message consists of one or more words, or bytes (8-bits). All messages conform to the protocol described below.

Table A-6: Message Protocol

⇓		Format Specifier, 0 = Single-word : 1 = Multiple Word
	⇓	Parameter Specifier, 0 for Multiple Word Message
X	**Y**	**OPCODE**
Number of Words to Follow (Multiword Message Only)		
Message Words (Multiword Message Only)		

Communications between the data monitor and the host are extremely error tolerant. If the Host sends a message that the data monitor doesn't understand, or doesn't conform to the message structure defined below, the data monitor will issue a protocol error and reset the communications interface causing no ill effects internal to the data monitor.

If a partial message is sent to the data monitor, the data monitor will time out after 200 milliseconds, issue a protocol error, and reset the communications interface causing no ill effects internal to the data monitor. Note that all of the data is passed to and from the data monitor in hexadecimal format.

There are two types of messages supported by the data monitor, *single-word* and *multiple word* messages. Each type of message is described in detail below.

Single/Multiple Word Messages

Single-word messages are defined by a *format specifier* of 0. These messages are a single word in length. Any parameters or options are specified by the *parameter specifier*.

Multiple word messages are defined by a format specifier of 1. These messages are variable in length. A multiple word message can be from 1 to 255 words long, or in the case of the download message, an indeterminate length.

Table A-7: Data Monitor Supported Messages

Message	Op-Code	Originator	Description
Set Storage Mode	0	Host	A single-word message updating the data monitor's Storage mode.

Message	Op-Code	Originator	Description
Read Storage Mode	32	Host	A single-word message requesting the data monitor's current Storage mode.
Storage Mode	32	Data Monitor	A single-word message reporting the data monitor's current Storage mode.
Set Average	1	Host	A multiword message updating the data monitor's Averaging mode.
Read Average	33	Host	A single-word message requesting the data monitor's current Averaging mode.
Average	33	Data Monitor	A multiword message reporting the data monitor's current Averaging mode.
Set Sample Inputs	2	Host	A multiword message updating the data monitor's inputs used for sampling.
Read Sample Inputs	34	Host	A single-word message requesting the data monitor's current sample inputs.
Sample Inputs	34	Data Monitor	A multiword message reporting the data monitor's current sample inputs.
Set Trigger Selects	3	Host	A multiword message updating the data monitor's trigger selects.
Read Trigger Selects	35	Host	A single-word message requesting the data monitor's current trigger selects.
Trigger Selects	35	Data Monitor	A multiword message reporting the data monitor's current trigger selects.
Set Trigger Data	4	Host	A multiword message updating the data monitor's trigger data.
Read Trigger Data	36	Host	A single-word message requesting the data monitor's current trigger data.
Trigger Data	36	Data Monitor	A multiword message reporting the data monitor's current trigger data.
Set User Time Base	5	Host	A multiword message updating the data monitor's current user time base.

Message	Op-Code	Originator	Description
Read User Time Base	37	Host	A single-word message requesting the data monitor's current user time base.
User Time Base	37	Data Monitor	A multiword message reporting the data monitor's current user time base.
Read System Voltage	38	Host	A single-word message requesting the data monitor's current system voltage.
System Voltage	38	Data Monitor	A multiword message reporting the data monitor's current system voltage.
Read Current Values	39	Host	A single-word message requesting the data monitor's current analog and digital input values
Current Values	39	Data Monitor	A multiword message reporting the data monitor's current analog and digital input values
Set Auto Download	0	Host	A multiword message updating the data monitor's current Auto Download status and period.
Read Auto Download	40	Host	A single-word message requesting the data monitor's current auto download status
Auto Download Status	40	Data Monitor	A multiword message reporting the data monitor's current Auto Download status.
Download EEPROM to Host	48	Host	A single-word message requesting the data monitor to download the EEPROMs current contents to the host.
Remote arm data monitor	61	Host	A single-word message requesting the data monitor to begin it's sample and store cycle.
Remote arm Response	61	Data Monitor	A single-word message reporting the results of the Remote arm data monitor message.
Re-initialize data monitor	62	Host	A multiword message re-initializing the data monitor with the current state.
Initialize data monitor	63	Host	A multiword message initializing and arming the data monitor.

Message	Op-Code	Originator	Description
Download Data	48	Data Monitor	A multiword message (up to 2k) reporting the data monitor's current EEPROM contents
Protocol Alarm	63	Data Monitor	A single-word message indicating an error has occurred in the data monitor.

Set Storage Mode

The Set Storage mode single-word message updates the data monitor's storage mode. This message is sent from the host to the data monitor. The Storage mode can be set to Store Immediate (1) or Store on Trigger (0). Store Immediate will use a trigger condition to begin data storage only if one is defined. Otherwise, the data monitor will sample and store data based solely on the user's time base. This mode makes the most of the data monitor's EEPROM since it doesn't have to store the time between stores; it is a fixed constant. Store on trigger will sample the inputs as fast as possible (~ 25 milliseconds) and store only those which satisfy the Trigger condition. By selective use of this mode bit and the data monitor's trigger, various storage modes can be achieved:

   ↳ *Store Immediate and no triggers defined:* The data monitor will store the active inputs based on the user's time base.

   ↳ *Store Immediate and triggers defined:* The data monitor will monitor all inputs until the trigger mode is satisfied. It will then store the trigger condition (and associated elapsed time). The data monitor will then fill the EEPROM based on the user's time base (no trigger).

   ↳ *Store on Trigger and triggers defined:* The data monitor will monitor all inputs until the trigger mode is satisfied. It will then store the trigger condition (and associated elapsed time). The data monitor will then continue monitoring the inputs and storing data only when the trigger condition is met.

✎ *Store on Trigger and no triggers defined:* This is a slightly illegal mode; however, the data monitor will still fill the EEPROM as follows. Because the data monitor has been tasked to store data on trigger but no triggers have been defined, the data monitor will default to storing the active inputs based solely on the user's time base.

The format of the Set Storage Mode message is defined in Tables A-8 and A-9. Bit 6 defines the storage/trigger mode.

Table A-8: Set Storage Mode Message

Word #	Bit 7	Bit 6	Bit 5	Bit 4	Bit 3	Bit 2	Bit 1	Bit 0
1	0	X	0	0	0	0	0	0

Table A-9: Storage Mode Definition

Bit # 6	Selected Storage Mode
0	Store on Trigger
1	Store Immediate

Read Storage Mode

The Read Storage mode single-word message requests the data monitor to report the Storage mode. This message is sent from the Host to the data monitor. The format of the Read Storage Mode message is defined in Table A-10.

Table A-10: Read Storage Mode Message

Word #	Bit 7	Bit 6	Bit 5	Bit 4	Bit 3	Bit 2	Bit 1	Bit 0
1	0	0	1	0	0	0	0	0

Storage Mode

The Storage Mode single-word message reports the data monitor's current Storage mode. This message is sent from the data monitor to the host. The Format of the Storage Mode message is defined in Tables A-11 and A-12. Bit 6 of the response defines the trigger mode.

Table A-11: Storage Mode

Word #	Bit 7	Bit 6	Bit 5	Bit 4	Bit 3	Bit 2	Bit 1	Bit 0
1	0	X	1	0	0	0	0	0

Table A-12: Storage Mode Bit Definition

Bit # 6	Selected Storage Mode
0	Store on Trigger
1	Store Immediate

Set Average Control

The Set Average multiword message updates the data monitor's analog input averaging. This message is sent from the host to the data monitor. The data monitor can be tasked to average all or none of the analog inputs. The averaging applies to all analog inputs and triggers, and cannot be enabled for some, and not for others. Once enabled, the data monitor will average all selected analog inputs and use this averaged value for storage AND for the trigger modes. The data monitor can average from one to eight (current plus last seven samples) samples. Bits 4, 3, 2 and 1 of word #2 contain the averaging selection word. 0x1 will average the current sample only, i.e. no averaging. 0x2 will average the current sample plus the previous average. 0x8 will average the current sample plus seven of the previous averages. 0x0 is an illegal average value and should not be used. Bit #0 enables or disables averaging.

The format of the Set Average control message is defined in Tables A-13 and A-14. Word #1 contains the OpCode. Word #2 contains the number of words to follow. Word #3 contains the averaging depth. Bits(4:1) contain the number of samples to average. Bit(0) enables or disables the averaging.

Table A-13: Average Control Message

Word #	Bit 7	Bit 6	Bit 5	Bit 4	Bit 3	Bit 2	Bit 1	Bit 0
1	1	0	0	0	0	0	0	1
2	0	0	0	0	0	0	0	1
3	0	0	0	AV3	AV2	AV1	AV0	Enable

Table A-14: Average Enable Definition

Bit # 0	Averaging
0	Average Disabled
1	Average Enabled

Read Average Control

The Read Average Control single-word message requests the data monitor to report the current Average Control mode. This message is sent from the host to the data monitor. The format of the Read Average Control message is defined in Table A-15.

Table A-15: Read Average Control

Word #	Bit 7	Bit 6	Bit 5	Bit 4	Bit 3	Bit 2	Bit 1	Bit 0
1	0	0	1	0	0	0	0	1

Average Control

The Average Control multiword message reports the data monitor's current Average Control Mode. This message is sent from the data monitor to the host. The format of the Average Control message is defined in Tables A-16 and A-17. Word #3 contains the current average information. The average depth is returned in Bit(4:1), and the average enable is returned in Bit(0).

Table A-16: Average Control Message

Word #	Bit 7	Bit 6	Bit 5	Bit 4	Bit 3	Bit 2	Bit 1	Bit 0
1	1	0	1	0	0	0	0	1
2	0	0	0	0	0	0	0	1
3	0	0	0	AV3	AV2	AV1	AV0	Enable

Table A-17: Average Enable Definition

Bit # 0	Averaging
0	Average Disabled
1	Average Enabled

Set Sample Inputs

The Set Sample Inputs multiword message updates the data monitor's sample inputs. This message is sent from the host to the data monitor. The data monitor can be tasked to store up to four digital inputs and four analog inputs. Writing a 1 in any bit location enables that input for storage. Writing a 0 in any bit location disables that input for storage. See Table A-18.

Table A-18: Set Sample Inputs Message

Word #	Bit 7	Bit 6	Bit 5	Bit 4	Bit 3	Bit 2	Bit 1	Bit 0
1	1	0	0	0	0	0	1	0
2	0	0	0	0	0	0	0	1
3	Analog 4	Analog 3	Analog 2	Analog 1	Digital 4	Digital 3	Digital 2	Digital 1

Read Sample Inputs

The Read Sample Inputs single-word message requests the data monitor to report the current selected Sample Inputs. This message is sent from the host to the data monitor. The format of the Read Sample Inputs message is defined in Table A-19.

Table A-19: Read Sample Inputs

Word #	Bit 7	Bit 6	Bit 5	Bit 4	Bit 3	Bit 2	Bit 1	Bit 0
1	0	0	1	0	0	0	1	0

Sample Inputs

The Sample Inputs multiword message reports the data monitor's sample inputs. This message is sent from the data monitor to the host. A logic high enables the selected input; a logic low disables the selected input, as shown in Table A-20.

Table A-20: Sample Inputs Message

Word #	Bit 7	Bit 6	Bit 5	Bit 4	Bit 3	Bit 2	Bit 1	Bit 0
1	0	0	1	0	0	0	1	0
2	0	0	0	0	0	0	0	1
3	Analog 4	Analog 3	Analog 2	Analog 1	Digital 4	Digital 3	Digital 2	Digital 1

Set Trigger Selects

The Set Trigger Selects multiword message updates the data monitor's trigger selects. This message is sent from the host to the data monitor. Words 3 through 10 define the triggers for each term. Refer to Table A-21 for details.

Table A-21: Set Trigger Selects Message

Word #	Bit 7	Bit 6	Bit 5	Bit 4	Bit 3	Bit 2	Bit 1	Bit 0
1	1	0	0	0	0	0	1	1
2	0	0	0	0	1	0	0	0
3	Trigger Select for Term 1							
4	Trigger Select for Term 2							
5	Trigger Select for Term 3							
6	Trigger Select for Term 4							
7	Trigger Select for Term 5							
8	Trigger Select for Term 6							
9	Trigger Select for Term 7							
10	Trigger Select for Term 8							

The format of the data monitor's trigger is shown below:

(Trigger_1 & Trigger_2 & Trigger_ 3) # (Trigger_4 & Trigger_5 & Trigger_6) # (Trigger_7 & Trigger_8)

Each of the eight triggers may choose any one of the four analog or four digital inputs, and one or more of the trigger modes specified as follows:

Digital Inputs

↳ Digital Input Rising

↳ Digital Input Falling

↳ Digital State (Level)

Analog Inputs

↳ Analog Input >Level

↳ Analog Input <= Level

↳ Analog Input Change >= or <= Threshold

For any given trigger, the user may specify one, two, or all of the trigger modes listed above. For example, to specify a trigger of toggle for a digital input, specify the rising and falling trigger modes. The format of the Trigger Control word is specified in Tables A-22 and A-23. Writing a logic 1 in the specified bit position enables the option. Writing a logic level 0 disables the option.

Table A-22: Digital Trigger Select Control Word

Bit	Digital Trigger Select Control Word
Bit 7	State (Level)
Bit 6	Input Falling
Bit 5	Input Rising
Bit 4	0 = Digital Inputs (constant)
Bit 3	Input = Input 3
Bit 2	Input = Input 2
Bit 1	Input = Input 1
Bit 0	Input = Input 0

Table A-23: Analog Trigger Select Control Word

Bit	Analog Trigger Selects Control Word
Bit 7	Input Change >= or <= Threshold
Bit 6	Input <= Value
Bit 5	Input >= Value
Bit 4	1 = Analog Inputs (constant)
Bit 3	Input = Input 3
Bit 2	Input = Input 2
Bit 1	Input = Input 1
Bit 0	Input = Input 0

Note in the above tables that multiple trigger modes may be selected (Bits 7, 6, or 5); however, only one input may be selected for each trigger (Bits 3, 2, 1, or 0).

For the trigger modes of Analog >=, Analog <=, and Analog Change, another word is needed to specify the thresholds or levels. This word is passed in the Trigger Data word. If a trigger does not need this data word, it should be left cleared (0x00). Place the HEX value into the corresponding trigger data word during the initialize command.

For the trigger mode of Digital State (level), the corresponding trigger data word should be set to 0xFF for a state of 1, and 0x00 for a state of zero.

Any combination of the eight trigger terms can be used. Unused terms in the trigger equation should have the Trigger Control Word and Trigger Data Word both set to 0x00.

Read Trigger Selects

The Read Trigger Selects single-word message requests the data monitor to report the current selected trigger selects. This message is sent from the host to the data monitor. The format of the Read Trigger Selects message is defined in Table A-24.

Table A-24: Read Trigger Selects

Word #	Bit 7	Bit 6	Bit 5	Bit 4	Bit 3	Bit 2	Bit 1	Bit 0
1	0	0	1	0	0	0	1	1

Trigger Selects

The Trigger Selects multiword message reports the data monitor's current trigger selects. This message is sent from the data monitor to the host. Table A-25 defines this.

Table A-25: Trigger Selects Message

Word #	Bit 7	Bit 6	Bit 5	Bit 4	Bit 3	Bit 2	Bit 1	Bit 0
1	1	0	1	0	0	0	1	1
2	0	0	0	0	1	0	0	0
3	Trigger Select for Term 1							
4	Trigger Select for Term 2							
5	Trigger Select for Term 3							
6	Trigger Select for Term 4							
7	Trigger Select for Term 5							
8	Trigger Select for Term 6							
9	Trigger Select for Term 7							
10	Trigger Select for Term 8							

Set Trigger Data

The Set Trigger Data multiword message updates the data monitor's trigger data. This message is sent from the host to the data monitor. Words 3 through 10 define the data for each term, as shown in Table A-26.

Table A-26: Set Trigger Data Message

Word #	Bit 7	Bit 6	Bit 5	Bit 4	Bit 3	Bit 2	Bit I	Bit 0
1	1	0	0	0	0	1	0	0
2	0	0	0	0	1	0	0	0
3	Trigger Data for Term 1							
4	Trigger Data for Term 2							
5	Trigger Data for Term 3							
6	Trigger Data for Term 4							
7	Trigger Data for Term 5							
8	Trigger Data for Term 6							
9	Trigger Data for Term 7							
10	Trigger Data for Term 8							

For the trigger modes of Analog >=, Analog <=, and Analog Change, another word is needed to specify the thresholds or levels. This word is passed in the Trigger Data word. If a trigger does not need this data word, it should be left cleared (0x00). Place the HEX value into the corresponding trigger data word during the initialize command.

For the trigger mode of Digital State (level), the corresponding trigger data word should be set to 0xFF for a state of 1, and 0x00 for a state of zero.

Any combination of the eight trigger terms can be used. Unused terms in the trigger equation should have the Trigger Control Word and Trigger Data Word both set to 0x00.

Read Trigger Data

The Read Trigger Data single-word message requests the data monitor to report the current selected trigger data. This message is sent from the host to the data monitor. The format of the Read Trigger Data message is defined in Table A-27.

Table A-27: Read Trigger Data

Word #	Bit 7	Bit 6	Bit 5	Bit 4	Bit 3	Bit 2	Bit I	Bit 0
1	0	0	1	0	0	1	0	0

Trigger Data

The Trigger Data multiword message reports the data monitor's current trigger data. This message is sent from the data monitor to the host, as shown in Table A-28.

Table A-28: Trigger Data Message

Word #	Bit 7	Bit 6	Bit 5	Bit 4	Bit 3	Bit 2	Bit 1	Bit 0
1	1	0	1	0	0	1	0	0
2	0	0	0	0	1	0	0	0
3	Trigger Data for Term 1							
4	Trigger Data for Term 2							
5	Trigger Data for Term 3							
6	Trigger Data for Term 4							
7	Trigger Data for Term 5							
8	Trigger Data for Term 6							
9	Trigger Data for Term 7							
10	Trigger Data for Term 8							

Read Current Values

The Read Current Values single-word message requests the data monitor to sample all four analog and all four digital inputs and report the results back to the host. This message is sent from the host to the data monitor. The format of the Read Current Values message is shown in Table A-29.

Table A-29: Read Current Values Message

Word #	Bit 7	Bit 6	Bit 5	Bit 4	Bit 3	Bit 2	Bit 1	Bit 0
1	0	0	1	0	0	1	1	1

Current Values

The Current Values multiword message reports the current state of the data monitor's four analog and all four digital inputs. This message is the result of the Read Current Values message. This message is sent from the data monitor to the host. The digital values reported shall be 0 or 1. The analog values reported shall range from 0 to 255 decimal. The format of the Current Values message is shown in Table A-30.

Table A-30: Current Values Message

Word #	Bit 7	Bit 6	Bit 5	Bit 4	Bit 3	Bit 2	Bit 1	Bit 0
1	1	0	1	0	0	1	1	0
2	0	0	0	0	0	1	0	1
3	0	0	0	0	Dig 4	Dig 3	Dig 2	Dig 1
4	Analog 1							
5	Analog 2							
6	Analog 3							
7	Analog 4							

Set Auto Download

The Auto Download multiword message updates the data monitor's Auto Download state. The Auto Download mode is used to configure the data monitor to transmit its stored data at a fixed interval without prompting by the host. This message is sent from the host to the data monitor. See Table A-31 for details.

Table A-31: Set Auto Download Message

Word #	Bit 7	Bit 6	Bit 5	Bit 4	Bit 3	Bit 2	Bit 1	Bit 0
1	1	0	0	0	0	1	1	0
2	0	0	0	0	0	1	0	1
3	Auto Download Enable/Disable							
4	Download Time LSB (milliseconds)							
5	Download Time							
6	Download Time							
7	Download Time MSB							

Byte #3 determines whether to enable or disable the auto download feature. A 0x00 in this byte will disable the auto download. A 0x01 will enable the auto download. Bytes 4 through 7 contain the desired auto download time in milliseconds. The minimum value is 5000 milliseconds; the maximum value is 4,294,967,295 milliseconds (0xFFFFFFFF).

Read Auto Download

The Read Auto Download single-word message requests the data monitor to report the current auto download state and period. This message is sent from the host to the data monitor. The format of the read auto download message is defined in Table A-32.

Table A-32: Read User Time Base

Word #	Bit 7	Bit 6	Bit 5	Bit 4	Bit 3	Bit 2	Bit 1	Bit 0
1	0	0	1	0	1	0	0	0

Auto Download Status

The Auto Download Status multiword message reports the data monitor's current auto download state and period. This message is sent from the data monitor to the host. Table A-33 describes this process.

Table A-33: Auto Download Status

Word #	Bit 7	Bit 6	Bit 5	Bit 4	Bit 3	Bit 2	Bit 1	Bit 0
1	1	0	0	0	0	1	1	0
2	0	0	0	0	0	1	0	1
3	Auto Download Enable/Disable							
4	Download Time LSB (milliseconds)							
5	Download Time							
6	Download Time							
7	Download Time MSB							

Byte #3 reports whether the auto download feature is enabled or disabled. A 0x00 in this byte will disable the auto download. A 0x01 will enable the auto download. Bytes 4 through 7 contain the desired auto download time in milliseconds. The minimum value is 5000 milliseconds. The maximum value is 4,294,967,295 milliseconds (0xFFFFFFFF).

Set User Time Base

The Set User Time Base multiword message updates the data monitor's user time base. The user time base is used to define the storage time in milliseconds between digital and analog samples. This message is sent from the host to the data monitor. Words 3 through 6 define the user time base in milliseconds. Convert the desired time base into a hexadecimal number and break over the four-word message. Refer to Table A-34 for details.

For example, a download of (0x00 0x11 0xFF 0X23) for bytes 3, 4, 5, and 6 respectively would select a time between samples of 0x0011FF23, or 1,179,427 milliseconds, or 1179.427 seconds, or 19.66 minutes.

Table A-34: Set User Time Base Message

Word #	Bit 7	Bit 6	Bit 5	Bit 4	Bit 3	Bit 2	Bit 1	Bit 0
1	1	0	0	0	0	1	0	1
2	0	0	0	0	0	1	0	0
3	User Time Base MSB (milliseconds)							
4	User Time Base							
5	User Time Base							
6	User Time Base LSB							

Read User Time Base

The Read User Time Base single-word message requests the data monitor to report the current user time base. This message is sent from the host to the data monitor. The format of the Read User Time Base message is defined in Table A-35.

Table A-35: Read User Time Base

Word #	Bit 7	Bit 6	Bit 5	Bit 4	Bit 3	Bit 2	Bit I	Bit 0
1	0	0	1	0	0	1	0	1

User Time Base

The User Time Base multiword message reports the data monitor's current user time base. This message is sent from the data monitor to the host. This is defined in Table A-36.

For example, a download of (0x00 0x11 0xFF 0X23) for bytes 3, 4, 5, and 6 respectively would select a time between samples of 0x0011FF23, or 1,179,427 milliseconds, or 1179.427 seconds, or 19.66 minutes.

Table A-36: User Time Base

Word #	Bit 7	Bit 6	Bit 5	Bit 4	Bit 3	Bit 2	Bit I	Bit 0
1	1	0	1	0	0	1	0	1
2	0	0	0	0	0	1	0	0
3	User Time Base MSB (milliseconds)							
4	User Time Base							
5	User Time Base							
6	User Time Base LSB							

Read System Voltage

The Read System Voltage single-word message requests the data monitor to report the current system voltage in tenths of volts. This message is sent from the host to the data monitor. The format of the read system voltage message is defined in Table A-37.

This message can be used to check the battery life of the data monitor. For example, if the data monitor returned a value of 85, this would indicate a battery voltage of 8.5 volts.

Table A-37: Read System Voltage

Word #	Bit 7	Bit 6	Bit 5	Bit 4	Bit 3	Bit 2	Bit 1	Bit 0
1	0	0	1	0	0	1	1	0

System Voltage

The System Voltage multiword message reports the data monitor's current system voltage in tenths of volts. This message is sent from the data monitor to the host. See Table A-38.

For example, if the data monitor returned a value of 85, this would indicate a battery voltage of 8.5 volts.

Table A-38: System Voltage

Word #	Bit 7	Bit 6	Bit 5	Bit 4	Bit 3	Bit 2	Bit 1	Bit 0
1	1	0	1	0	0	1	1	0
2	0	0	0	0	0	0	0	1
3	System Voltage (tenths of volts)							

Download EEPROM to Host

The Download EEPROM to Host single-word message requests the data monitor to download the EEPROM contents to the host. This message is sent from the host to the data monitor. The Format of the Download EEPROM to Host message is defined in Table A-39. The data monitor will send all valid data stored after the last arm command upon receipt of this message.

Table A-39: Download EEPROM to Host

Word #	Bit 7	Bit 6	Bit 5	Bit 4	Bit 3	Bit 2	Bit 1	Bit 0
1	0	0	1	1	0	0	0	0

Download Data

The Download Data multiword message reports the data monitor's current EEPROM contents. This message is sent from the data monitor to the host. Due to the variability and depth of the EEPROM contents, this message is of an indeterminate length up to 2 kilobytes. The data monitor shall set the number of words to follow to 0xFF specifying an indeterminate length message. The host should accept data until the data monitor stops sending data. This process is defined in Table A-40.

Table A-40: Download Data

Word #	Bit 7	Bit 6	Bit 5	Bit 4	Bit 3	Bit 2	Bit I	Bit 0
1	1	0	1	1	0	0	0	0
2	1	1	1	1	1	1	1	1
3 - N	Download Data, Varying Length							

Initialize Data Monitor

The Initialize Data Monitor multiword message initializes and arms the data monitor. Once completed, a simple press of the status button will arm the unit beginning the sampling cycle. The format of the Initialize message is shown in Table A-41.

Table A-41: Initialize Data Monitor

Initialize Word	Description
1	0xBF
2	0x1D
3	Control/Average Word
4	Store Enable Word
5	Trigger Control 1
6	Trigger Data 1
7	Trigger Control 2
8	Trigger Data 2
9	Trigger Control 3
10	Trigger Data 3

Initialize Word	Description
11	Trigger Control 4
12	Trigger Data 4
13	Trigger Control 5
14	Trigger Data 5
15	Trigger Control 6
16	Trigger Data 6
17	Trigger Control 7
18	Trigger Data 7
19	Trigger Control 8
20	Trigger Data 8
21	User Count MSB
22	User Count
23	User Count
24	User Count LSB
25	Time Stamp MSB
26	Time Stamp
27	Time Stamp
28	Time Stamp LSB
29	Date Stamp (Day)
30	Date Stamp (Month)
31	Date Stamp (Year)

Re-initialize Data Monitor

The Re-initialize Data Monitor multiword message re-initializes the data monitor with the parameters previously set by the Initialize Data Monitor command. This message is sent from the host to the data monitor.

This message provides a quick way to restart the current sample cycle. After the Re-initialize has been issued, a press of the arm/status button, or a Remote Arm message, will begin the sample and store cycle.

Note that all parameters are lost if the power is removed from the data monitor. Issuing a Re-initialize message after power has been cycled will result in inconsistent behavior from the data monitor. Table A-42 defines this process.

Table A-42: Re-Initialize data monitor

Word #	Bit 7	Bit 6	Bit 5	Bit 4	Bit 3	Bit 2	Bit I	Bit 0
1	1	0	1	1	1	1	1	0
2	0	0	0	0	0	1	0	0
3	Time Stamp MSB (Hour)							
4	Time Stamp (Minute)							
5	Time Stamp (Second)							
6	Time Stamp LSB (100th Second)							
7	Date Stamp (Day)							
8	Date Stamp (Month)							
9	Date Stamp (Year)							

Remote Arm

The Remote Arm single-word message is used by the host to arm the data monitor without having to press the Arm/Status button. This message is sent from the host to the data monitor. The data monitor must be initialized before the Remote Arm message is issued, or an error will result. The format of the Remote Arm message is shown in Table A-43.

Table A-43: Remote Arm Message

Word #	Bit 7	Bit 6	Bit 5	Bit 4	Bit 3	Bit 2	Bit I	Bit 0
1	0	0	1	1	1	1	0	1

Remote Arm Response

The host issues the Remote Arm Response single-word message after a remote arm has been issued. This message is sent from the data monitor to the host. The format of the Remote Arm Response message is shown in Table A-44 and the remote arm results are in Table A-45.

Table A-44: Remote Arm Response Message

Word #	Bit 7	Bit 6	Bit 5	Bit 4	Bit 3	Bit 2	Bit 1	Bit 0
1	0	X	1	1	1	1	0	1

Table A-45: Remote Arm Results

Bit # 6	Remote Arm Status
0	Remote Arm Failed
1	Remote Arm Successful

Bit #6 of byte #1 specifies whether or not the Remote Arm message was successful. A 0 indicates the Remote Arm was not successful. A 1 indicates that the Remote Arm was successful.

A Remote Arm will fail if the data monitor has not been initialized before the Remote Arm message was sent.

Protocol Alarm

The Protocol Alarm single-word message is used by the data monitor to let the host know there has been a communication error. The data monitor sends the Protocol alarm to the host. The various protocol errors are listed below. Bit #6 defines the Protocol Alarm type. See Tables A-46 and A-47.

Table A-46: Protocol Alarm Message

Word #	Bit 7	Bit 6	Bit 5	Bit 4	Bit 3	Bit 2	Bit 1	Bit 0
1	0	X	1	1	1	1	1	1

Table A-47: Protocol Alarm Definition

Bit # 6	Alarm Type
0	Host Time Out
1	Unused OpCode

Firmware Flow Overview

The data monitor's PIC firmware can be broken into two main components: An interrupt service routine where timing critical events are processed, and a main loop where nontiming critical events occur. Figures A-15 and A-16 provide simplified flow diagrams for these routines.

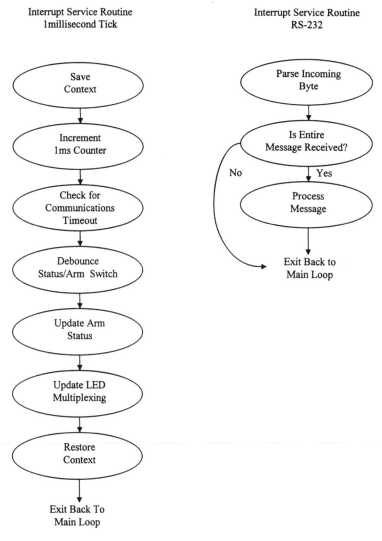

Figure A-15: Interrupt service routine flow diagram

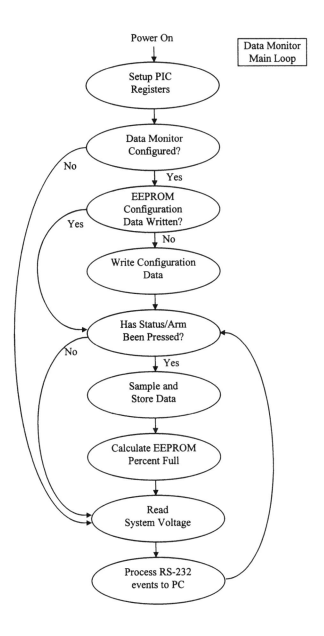

Figure A-16: Main loop flow diagram

The main loop performs all of the non-timing critical operations. When power is first applied to the data monitor, the PIC initializes all of its internal variables and waits for the user to configure the data monitor, and then for the user to press the Arm button. Until this point, the only thing monitored by the data monitor is the system raw voltage. This allows the user to check the system voltage at any time. When armed, the data monitor writes the configuration parameters to the EEPROM, and begins sampling and storing data based on the user-supplied parameters. Once sampling has begun, the PIC also calculates the EEPROM percent full, and sends any RS-232 data that may be pending over the COM port to the PC.

The interrupt service routine, or ISR, is broken into two main sections. One section services the 1 millisecond tick, or timer, used to run the internal real time clock. This means that all events inside the ISR must be finished in less than 1 millisecond; otherwise, the next tick will be missed, thus corrupting the real time clock. The other section services the RS-232 receive interrupt. This routine services all data received from the PC to the data monitor over the COM port.

The first thing that must happen in any ISR is to save the *context*. A context is nothing more than a snapshot of all important registers, like the working or status registers, so that they may be restored when servicing of the interrupt is complete. As an example, suppose a routine was running that had just subtracted two numbers. It was about to examine the zero bit of the status register to see if the two numbers were equal (if you subtract two numbers and the result is zero, they were equal). Just as the zero bit of the status register was about to be tested, an interrupt comes along and we jump off to service it. Inside the ISR, we service the interrupt which, as a side effect, changes the zero bit of the status register, and then return to where we jumped out. Now, since the zero bit of the status register has been corrupted, the main loop now does the wrong thing. Saving the context when we enter the ISR and restoring it when complete solves problems of this type.

The 1-millisecond ISR performs all of the time critical functions of the data monitor. The first operation it performs is to update a 1-millisecond counter. This counter is the real time clock for the data monitor, and represents the smallest amount of time the data monitor can recognize. Next, this routine checks to make sure that communications from the PC to the data monitor are okay. It does this by only allowing a certain amount of time for each message to complete. If a message takes too long, it is aborted and all queues are flushed. This insures that if communications ever get out of sync, all we have to do is wait and the error will correct itself beginning with the next message. The time-out period is set to 200 milliseconds. The next function performed by this routine is to monitor the status/arm switch. This switch is also debounced in the firmware. If the status/arm switch is pressed and the unit is configured but not armed, the PIC arms the data monitor to begin the sample and store cycle. The last function performed is to turn the LED display window on. If the status/arm button is pressed, the LED status windows must be enabled so that the user can see the status.

The RS-232 ISR receives all messages from the PC and processes them. Each message consists of one or more words, or bytes.

This routine waits for all of the bytes in a message to be received, processes the message, and sends the appropriate response back to the PC. The response bytes are placed in a queue which is monitored by the main loop. The main loop removes the bytes from the queue and sends them to the PC.

Initialize Routine

The supplied software Dmon.exe takes care of the initialize routine. However, for those who live for details, or want to write their own initialize routine, the details are given in Table A-48.

Table A-48: Initialize Byte Sequence

Initialize Byte	Description
1	Initialize Opcode
2	# Bytes to follow
3	Control/Average Word
4	Store Enable Word

Initialize Byte	Description
5	Trigger Control 1
6	Trigger Data 1
7	Trigger Control 2
8	Trigger Data 2
9	Trigger Control 3
10	Trigger Data 3
11	Trigger Control 4
12	Trigger Data 4
13	Trigger Control 5
14	Trigger Data 5
15	Trigger Control 6
16	Trigger Data 6
17	Trigger Control 7
18	Trigger Data 7
19	Trigger Control 8
20	Trigger Data 8
21	User Count MSB
22	User Count
23	User Count
24	User Count LSB
25	Time Stamp MSB
26	Time Stamp
27	Time Stamp
28	Time Stamp LSB
29	Date Stamp (Day)
30	Date Stamp (Month)
31	Date Stamp (Year)

Initialize Opcode

The Initialize Opcode is used to begin an initialize message. This message initializes the data monitor and readies it for recording. Once initialized, pressing the arm/status button can arm the data monitor. The data monitor will display 00 when armed. Subsequent presses of the status button will display the data monitor's memory used, in percent. The initialize Opcode is 0xBF.

Number of Bytes to Follow

This byte specifies the number of bytes in the message to follow, not including this byte. The number of bytes to follow for the initialize message is 30 or 0x1D.

Store/Average Word

The Store/Average word specifies operating parameters for the data monitor, as described in Table A-49.

Table 49: Store/Average Word

Bit	Mode Control Word
Bit 7	1 = Store Immediate / 0= Store on Trigger
Bit 6	Reserved (0)
Bit 5	Reserved (0)
Bit 4	Average Value, MSB
Bit 3	Average Value
Bit 2	Average Value
Bit 1	Average Value, LSB
Bit 0	Average Enable (1=enabled)

Selecting Store Immediate (1) will use a trigger condition to begin data storage only if one is defined. Otherwise the data monitor will sample and store data based solely on the user's time base. This mode makes the most of the data monitor's EEPROM since it doesn't have to store the time between stores—it is a fixed constant.

Store on Trigger (0) will sample the inputs as fast as possible (~ 25 milliseconds) and store only those which satisfy the trigger condition. When the trigger is satisfied, the data monitor will store the active inputs along with the elapsed time since the last successful trigger condition. Selective use of this mode bit and the data monitor's trigger can achieve various storage modes.

The various trigger modes are defined as follows:

✎ *Store Immediate and no triggers defined*: The data monitor will store the active inputs based on the user's time base.

✎ *Store Immediate and triggers defined:* The data monitor will monitor all inputs until the trigger mode is satisfied and then store the trigger condition (and associated elapsed time). The data monitor will then fill the EEPROM based on the user's time base (no trigger).

✎ *Store on Trigger and triggers defined:* The data monitor will monitor all inputs until the trigger mode is satisfied. It will then store the trigger condition (and associated elapsed time). The data monitor will then continue monitoring the inputs and storing data only when the trigger condition is met.

✎ *Store on Trigger and no triggers defined:* This is a slightly illegal mode; however, the data monitor will still fill the EEPROM. Because the data monitor has been tasked to store data on trigger but no triggers have been defined, the data monitor will default to storing the active inputs based solely on the user's time base.

The data monitor can also be tasked to average all or none of the analog inputs. Once enabled, the data monitor will average all selected analog inputs and use this averaged value for storage AND for the trigger modes. The data monitor can average from 1 to 8 (current plus last seven samples) samples. Bits 4, 3, 2 and 1 contain the averaging selection word. 0x1 will average the current sample. 0x2 will average the current sample plus the previous average. 0x8 will average the current sample plus seven of the previous averages. 0x0 is an illegal average value and should not be used.

Store Enable Word

The Store Enable Word is used to enable/disable the analog and digital inputs to the data monitor. There are four analog and four digital inputs, each of which can be separately enabled or disabled. All or any of the data monitor's inputs can be used in the trigger generation regardless of whether they are stored. This is defined in Table A-50.

Table A-50: Store Enable Word

Bit	Store Enable Word
Bit 0	1 enables Digital 1, 0 disables
Bit 1	1 enables Digital 2, 0 disables
Bit 2	1 enables Digital 3, 0 disables
Bit 3	1 enabled Digital 4, 0 disables
Bit 4	1 enables Analog 1, 0 disables
Bit 5	1 enables Analog 2, 0 disables
Bit 6	1 enables Analog 3, 0 disables
Bit 7	1 enables Analog 4, 0 disables

Trigger Control/Data Words

The Format of the data monitor's trigger is shown below:

(Trigger_1 & Trigger_2 & Trigger_ 3) # (Trigger_4 & Trigger_5 & Trigger_6) # (Trigger_7 & Trigger_8)

Each of the eight triggers may choose any one of the four analog or four digital inputs, and one or more of the trigger modes specified below.

Digital Inputs

↪ Digital Input Rising

↪ Digital Input Falling

↪ Digital State (Level)

Analog Inputs

♻ Analog Input >=Level

♻ Analog Input <= Level

♻ Analog Input Change >= or <= Threshold

For any given trigger, the user may specify one, two, or all of the trigger modes listed above. For example, to specify a trigger of toggle for a digital input, specify the rising and falling trigger modes. The format of the Trigger Control byte is specified below. Writing a logic1 in the specified bit position enables the option; writing a logic level 0 disables the option. These triggers are defined in Tables A-51 and A-52:

Table A-51: Digital Trigger Control Word

Bit	Digital Trigger Control Word
Bit 7	State (Level)
Bit 6	Input Falling
Bit 5	Input Rising
Bit 4	0 = Digital Inputs
Bit 3	Input = Input 3
Bit 2	Input = Input 2
Bit 1	Input = Input 1
Bit 0	Input = Input 0

Table A-52: Analog Trigger Control Word

Bit	Analog Trigger Control Word
Bit 7	Input Change <> Threshold
Bit 6	Input <= Value
Bit 5	Input >= Value
Bit 4	1 = Analog Inputs
Bit 3	Input = Input 3
Bit 2	Input = Input 2
Bit 1	Input = Input 1
Bit 0	Input = Input 0

Note in the above tables that multiple trigger modes may be selected (Bits 7, 6, or 5). However, only one input may be selected for each trigger (Bits 3, 2, 1, or 0).

For the trigger modes of Analog >=, Analog <=, and Analog Change, another byte is needed to specify the thresholds or levels. This byte is passed in the Trigger Data word. If a trigger does not need this data word, it should be left cleared (0x00). Place the HEX value into the corresponding trigger data word during the initialize command.

For the trigger mode of Digital State (level), the corresponding trigger data word should be set to 0xFF for a state of 1 and 0x00 for a state of zero.

Any combination of the eight trigger terms can be used. Unused terms in the trigger equation should have the Trigger Control Word and Trigger Data Word both set to 0x00.

User Time Base

The User Time Base bytes are used to specify the time in milliseconds between samples. This value is specified in hexadecimal. This value is only used when a trigger condition is not active. In trigger mode, these bytes are ignored. When the User Time Base has expired, all selected inputs are sampled. In trigger mode, all inputs are sampled at full speed (approximately 25 milliseconds) and stored only when the trigger conditions are satisfied.

For example, a download of (0x00 0x11 0xFF 0X23) for bytes 21, 22, 23, and 24 respectively would select a time between samples of 0x0011FF23, or 1,179,427 milliseconds, or 1179.427 seconds, or 19.66 minutes.

Time Stamp Bytes

The Time Stamp bytes are used to specify the current time to the nearest second to the data monitor. These bytes are stored in EEPROM and used during data recovery to quantify the stored data. All values are specified in hexadecimal and in 24-hour format. The format of the Time Stamp bytes is shown in Table A-53.

Table A-53: Time Stamp Bytes

Byte	Time Stamp Bytes
1	Hour Byte
2	Minute Byte
2	Second Byte
3	100th Second Byte

For example, to specify a time of 4:23:10.02pm, or 16:23:10.02, the Time Bytes would be as follows:

0x10 0x17 0x0A 0x02 for bytes 25, 26, 27, and 28 respectively.

Date Stamp Bytes

The date stamp bytes are used to specify the current date in MM/DD/YY format to the data monitor. These bytes are stored in EEPROM and used during data recovery to quantify the stored data. All values are specified in hexadecimal. The format of the Date Stamp bytes is shown in Table A-54.

Table A-54: Date Stamp Bytes

Byte	Date Stamp Bytes
1	Day byte (0-31)
3	Month byte (0-12)
4	Year byte (00-99)

For example, to specify a date of April 21, 1996, the Date Bytes would be as follows:

21 04 96 or 0x15 0x04 0x60 for bytes 29, 30, and 31 respectively.

EEPROM Storage Format

The data monitor contains a 16-kilobytes EEPROM (or 2 kilobytes, 8-bit words) used to store data during operation. Enough configuration data is stored in the EEPROM to insure recovery of the stored data even if the data monitor loses power during operation. The table entries marked with *'s indicate the EEPROM entries that are repeated until the EEPROM is full. Note that all data that is stored in the EEPROM and downloaded to the host is in hexadecimal format.

Table A-55 gives a memory "snapshot" for Store Immediate with no triggers defined and Store on Trigger with no triggers defined, while Table A-56 gives similar information for Store Immediate and Triggers Defined. Table A-57 does the same for Store on Trigger and Triggers Defined.

Table A-55: EEPROM Format Type 1

Byte Address	Byte Value	Description
0x00	Address	Address of last modified byte
0x01	Page	Page of last modified byte
0x02	Control Average	Control / Average Word
0x03	Store Enables	Selected Digital and Analog inputs
0x04	Time MSB	Time Stamp, hour byte
0x05	Time	Time Stamp, minute byte
0x06	Time	Time Stamp, second byte
0x07	Time LSB	Time Stamp, 100th second byte
0x08	Date (Day)	Date Stamp, Day byte
0x09	Date (Month)	Date Stamp, Month byte
0x0A	Date (Year)	Date Stamp, Year byte
0x0B	User Time MSB	Selected user time byte MSB
0x0C	User Time	Selected user time byte
0x0D	User Time	Selected user time byte
0x0E	User Time LSB	Selected user time byte LSB
0x0F	Store Offset MSB	Elapsed time from arm
0x10	Store Offset	Elapsed time
0x11	Store Offset	Elapsed time
0x12	Store Offset LSB	Elapsed time
*0x13	Digital Inputs	Digital Input byte (if selected, see text)
*0x14	Analog Input 1	Analog input 1 (if selected, see text)
*0x15	Analog Input 2	Analog input 2 (if selected, see text)
*0x16	Analog Input 3	Analog input 3 (if selected, see text)
*0x17	Analog Input 4	Analog input 4 (if selected, see text)

Table A-56: EEPROM Format Type 2

Byte Address	Byte Value	Description
0x00	Address	Address of last modified byte
0x01	Page	Page of last modified byte
0x02	Control Average	Control / Average Word
0x03	Store Enables	Selected Digital and Analog inputs
0x04	Time MSB	Time Stamp, hour byte
0x05	Time	Time Stamp, minute byte
0x06	Time	Time Stamp, second byte
0x07	Time LSB	Time Stamp, 100th second byte
0x08	Date (Day)	Date Stamp, Day byte
0x09	Date (Month)	Date Stamp, Month byte
0x0A	Date (Year)	Date Stamp, Year byte
0x0B	User Time MSB	Selected user time byte MSB
0x0C	User Time	Selected user time byte
0x0D	User Time	Selected user time byte
0x0E	User Time LSB	Selected user time byte LSB
0x0F	Store Offset MSB	Elapsed time from arm
0x10	Store Offset	Elapsed time
0x11	Store Offset	Elapsed time
0x12	Store Offset LSB	Elapsed time
0x13	Trigg offset MSB	Elapsed time from arm to first trigger
0x14	Store Offset	Elapsed time
0x15	Store Offset	Elapsed time
0x16	Trig Offset LSB	Elapsed time
*0x17	Digital Inputs	Digital Input byte (if selected, see text)
*0x18	Analog Input 1	Analog input 1 (if selected, see text)
*0x19	Analog Input 2	Analog input 2 (if selected, see text)
*0x1A	Analog Input 3	Analog input 3 (if selected, see text)
*0x1B	Analog Input 4	Analog input 4 (if selected, see text)

Table A-57: EEPROM Format Type 3

Byte Address	Byte Value	Description
0x00	Address	Address of last modified byte
0x01	Page	Page of last modified byte
0x02	Control Average	Control / Average Word
0x03	Store Enables	Selected Digital and Analog inputs
0x04	Time MSB	Time Stamp, hour byte
0x05	Time	Time Stamp, minute byte
0x06	Time	Time Stamp, second byte
0x07	Time LSB	Time Stamp, 100th second byte
0x08	Date (Day)	Date Stamp, Day byte
0x09	Date (Month)	Date Stamp, Month byte
0x0A	Date (Year)	Date Stamp, Year byte
0x0B	User Time MSB	Selected user time byte MSB
0x0C	User Time	Selected user time byte
0x0D	User Time	Selected user time byte
0x0E	User Time LSB	Selected user time byte LSB
0x0F	Arm Offset MSB	Elapsed time from arm
0x10	Arm Offset	Elapsed time
0x11	Arm Offset	Elapsed time
0x12	Arm Offset LSB	Elapsed time
*0x13	Trigg offset MSB	Elapsed time from arm to trigger
*0x14	Trig Offset	Elapsed time
*0x15	Trig Offset	Elapsed time
*0x16	Trig Offset LSB	Elapsed time
*0x17	Digital Inputs	Digital Input byte (if selected, see text)
*0x18	Analog Input 1	Analog input 1 (if selected, see text)
*0x19	Analog Input 2	Analog input 2 (if selected, see text)
*0x1A	Analog Input 3	Analog input 3 (if selected, see text)
*0x1B	Analog Input 4	Analog input 4 (if selected, see text)

EEPROM Address/Page

These two bytes record the last EEPROM address updated by the data monitor. The data monitor and host use these locations to recover data from the EEPROM. Even if the data monitor loses power, these bytes will insure that the most possible amount of user data can be recovered. The EEPROM is segmented into eight pages, each 256 bytes or words deep.

Control/Average Word

This byte records the control/average control word.

Store Enable Word

This byte records the selected digital and analog inputs that the data monitor used during recording.

Time Stamp

These bytes record the time at which the data monitor was initialized. This time plus the offset time plus the User Count time determines the exact time of the first data store.

Date Stamp

These bytes record the date at which the data monitor was initialized.

User Time Base

These bytes record the user's count bytes. This value specifies the time in milliseconds between samples.

Arm Offset Count

These bytes record the time in milliseconds between the initialize, or time stamp and the arm of the data monitor.

Trigger Offset Count

These bytes record the time in milliseconds between the arm or sample and the next valid sample. Samples are qualified by the defined trigger(s).

Digital/Analog Inputs

These bytes record the active digital and/or analog inputs whenever the trigger or user time base has been satisfied.

Data Monitor Schematics and PCB Foils

The next few sheets contain the artworks for the data monitor PCB and the General IO / Temperature PCB. Although both of these circuit boards are double-sided, there are only a few vias on the data monitor PCB. Most of the top-to-bottom connections are completed with the leg of a through-hole component. When assembling, make sure you solder both sides of the circuit board.

These foil patterns are also stored on the CD enclosed with this book. If you have access to a printer, it is recommended that you print the foil patterns directly. They are stored in Adobe PDF format. If you do not have Adobe Acrobat reader available on your PC, the freeware version is also available on the CD.

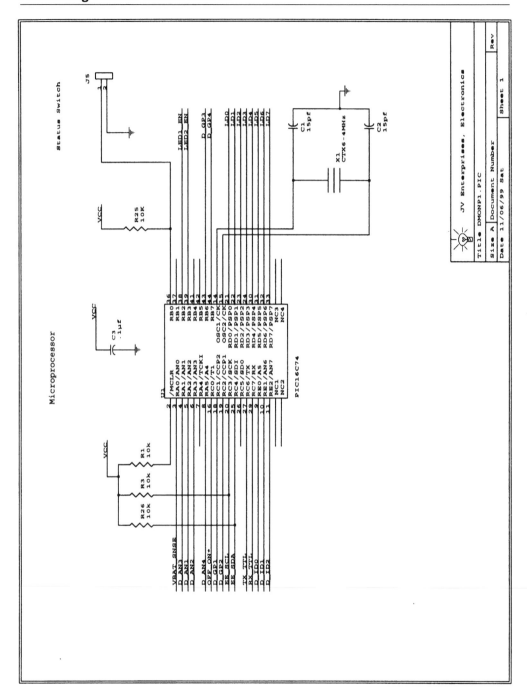

Figure A-17: Data Monitor schematics

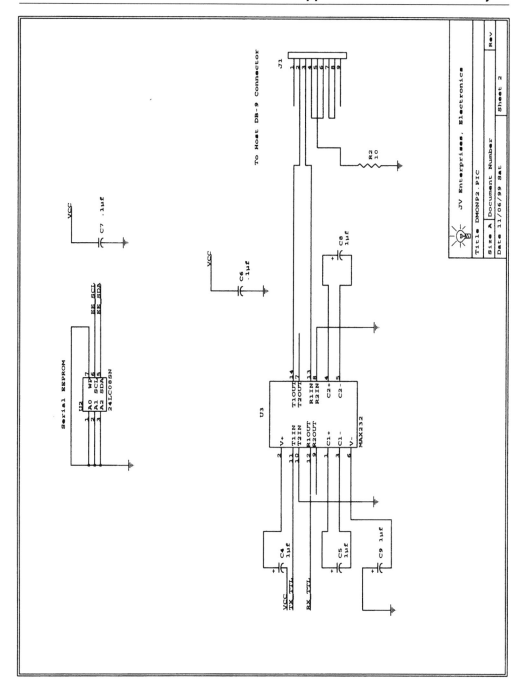

Figure A-18: Data Monitor schematics

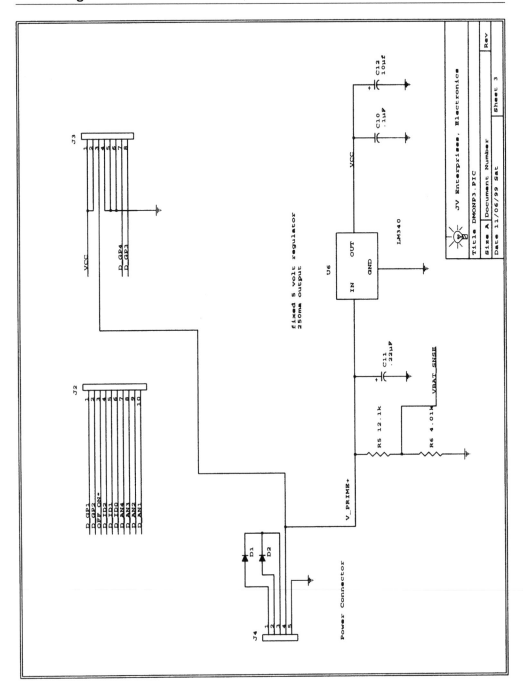

Figure A-19: Data Monitor schematics

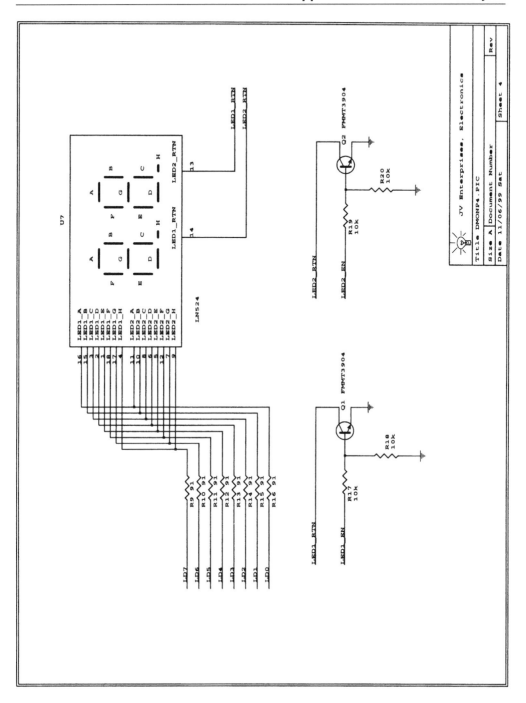

Figure A-20: Data Monitor schematics

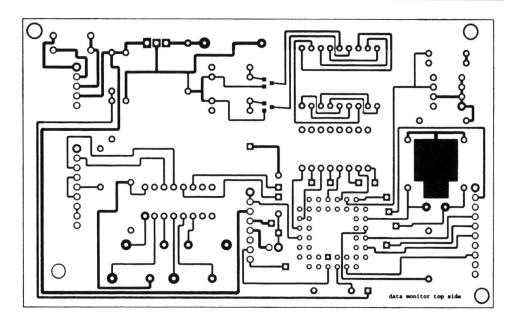

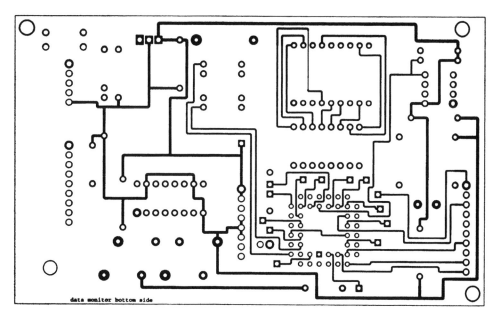

Figures A-21 and A-22: Data Monitor top and bottom foils

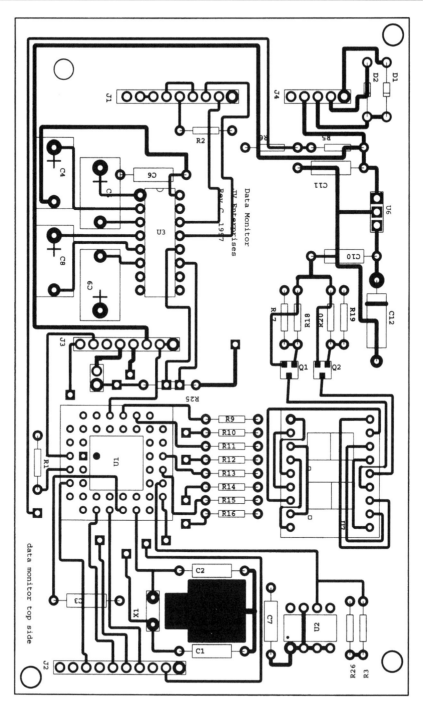

Figure A-23: Data Monitor assembly drawing

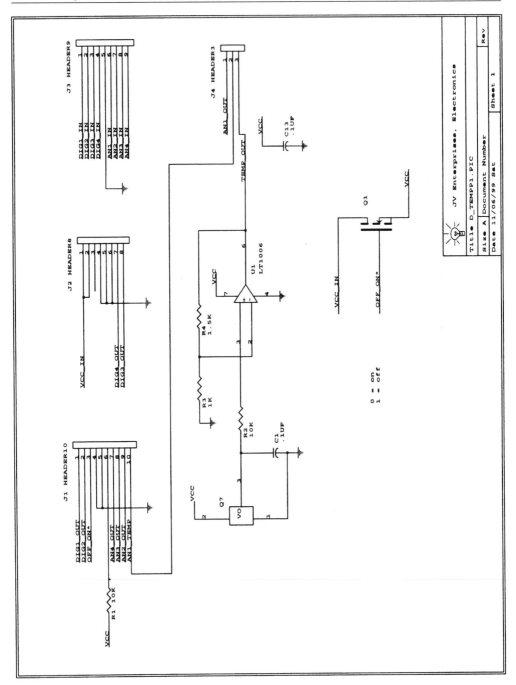

Figure A-24: General I/O-Temperature schematics

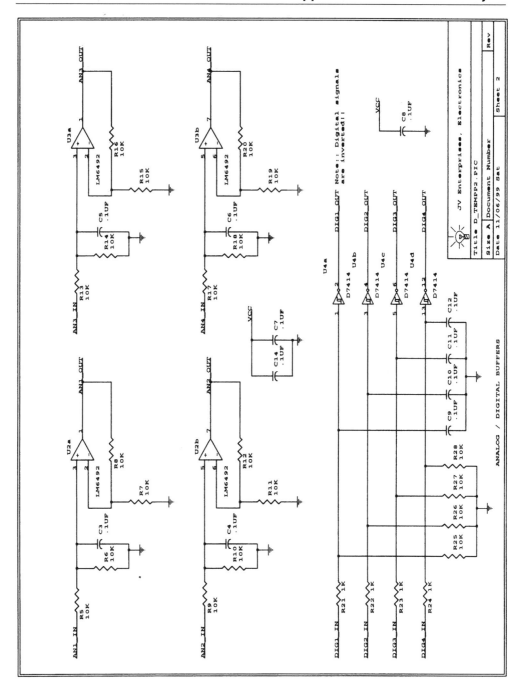

Figure A-25: General I/O-Temperature schematics

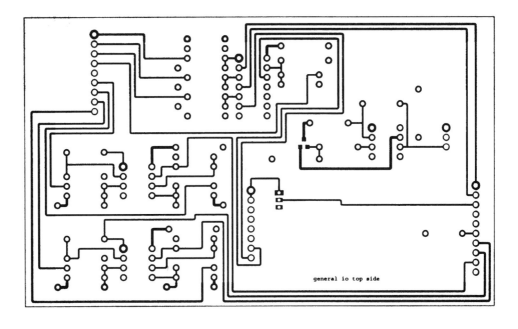

general io top side

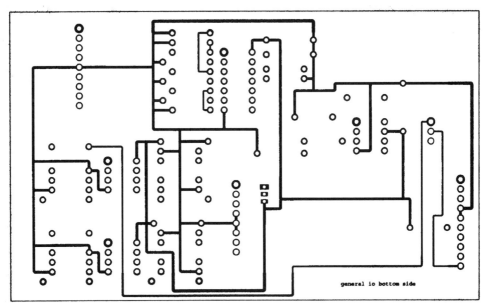

general io bottom side

Figures A-26 and A-27: General I/O-Temperature top and bottom foils

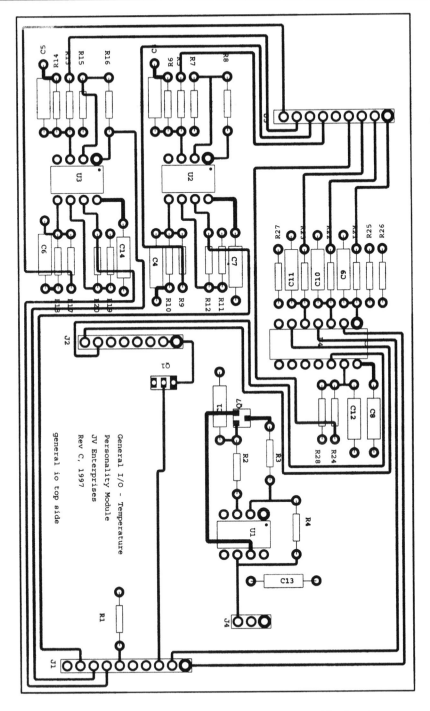

Figure A-28: General I/O-Temperature assembly drawing

Data Monitor Parts List

Main Circuit Board

✎ A complete kit of all components, PCBs, chassis and associated parts is available from JV enterprises. Check our web page for current prices, or just give us a call.

Quan	Part Number	Description	Ref Des
1	-	Data Monitor printed circuit board	
1	-	Solder, fine 6 feet	
1	-	Solder wick 1 spool	
1	SR272B	Black plastic enclosure with 9v battery compartment	
1	-	Hole Template Sticker	
1	-	Data Monitor front panel label	
1	-	120v AC -> 12v DC 300ma power brick	
1	AE1020	MF RS-232 Data Cable, 2M	
1	DM-SW1	Data Monitor software, source and executables	
1	929834-02-36	straight male , .1" center, .025" square post (36)	J2,3
1	929835-01-36	right angle male, .1" center, .025" square post (36)	J1,4,5
1	929974-01-36	straight female, .1" center, .025" square post (36)	
1	109F	9-pin D female connector	
1	109M	9-pin D male connector	
2	A9006	9-pin D mounting kit, female lock screws	
1	72K	9 volt battery connector 6" leads	
1	-	Multi-strand ribbon cable 2 feet	
1	-	22 gauge red wire, 3 feet	
1	-	22 gauge black wire, 2 feet	
1	-	1/16" heat shrink tubing 25 pieces	
1	-	1/8" heat shrink tubing 2 pieces	
4	J241	1/2" threaded standoffs	
4	-	4-40 panhead screws 1/4"	
4	-	4-40 flathead screws 1/4"	

2	1012PHCT	15pf 100v axial capacitor	C1,2
4	1109PHCT	.1uf, 50v axial capacitor	C3,6,7,10
1	1211PHCT	.22uf, 50v axial capactitor	C11
2	1N4001GICT	GP rectifier Diode, .5a forward current	D1,D2
1	24LC16B/P	16k EEPROM	U2
1	A417	44 pin PLCC solder tail socket	U1
1	CTX006-4MHZ	4mhz oscillator	X1
1	EG1502	Power switch, rocker	
1	EG2041	Arm switch, momemtary	
2	FMMT3904	NPN transistors, SOT-23	Q1,2
1	HR110	Power plug, male 4 pin (AC adaptor)	
1	HR213	Power receptacle, female 4 pin (chassis)	
1	LM340T	5volt regulator	U6
1	MAX232CPE	RS232 — TTL converter	U3
1	10QBK	10 1/4w axial resistor 5%	R2
8	10KQBK	10.0k 1/4w axial resistor 5%	R1,3,17,18 19,20,25,26
1	12.1KXBK	12.1k 1/4w axial resistor 1%	R5
1	4.02KXBK	4.02k 1/4w axial resistor 1%	R6
1	P355-ND	LN524 2 digit LED, common cathode	U7
8		91QBK 91 1/4w axial resistor 5%	R9,10,11,12 13,14,15,16
4	P5988	1uf 63v axial capacitor	C4,5,8,9
1	P6343	10uf, 25v electrolytic axial capacitor	C12
1	PIC16C74	PIC microprocessor, programmed	U1
1	PRD180B	Bezel, black	
1	PRD180C	Lens, 1.8" clear	

General IO / Temp Parts List

Quan	Part Number	Description	Ref Des
1	-	General IO / Temperature printed circuit board	
1	929835-01-36	right angle male, .1" center, .025" square post (36)	J3,4
1	929974-01-36	straight female, .1" center, .025" square post (36)	J1,2
1	109F	9-pin D female connector	
1	A1236	9-pin D male retainer screws	
1	A2051	9-pin D plastic housing	
1	929950	Shorting Jumper	
5	-	Test plug to test plug connectors, various colors	
1	-	Sheet of analog and digital connector labels	
1	-	1/16" heat shrink tubing 25 pieces	
13	1109PHCT	.1uf, 50v axial capacitor	C1,3,4,5,6,7, 8,9,10,11,12, 13,14
1	MM74HC14	Schmitt trigger inverter	U4
2	LM6492BEN	OP-AMP, GP	U2,3
1	LM50CIM3	Temperature sensor	Q7
1	LT1006CN8	OP-AMP, GP	U1
5	1.0KQBK	1.0k 1/4w axial resistor 5%	R3,21,22, 23,24
1	1.5KQBK	1.5k 1/4w axial resistor 5%	R4
22	10KQBK	10.0k 1/4w axial resistor 5%	R1,2,5,6,7,8, 9,10,11,12, 13,14,15,16, 17,18,19,20 25,26,27,28
1	ZVP2106A	P-channel MOSFET	Q1

APPENDIX B
Building the Data Monitor

To properly and safely construct the data monitor, you must possess certain skills and equipment. We'll summarize these before getting into the construction details.

Skills Needed

⤷ *Soldering skills.* These should be moderate to advanced. You should have assembled a few kits or projects successfully and have experience in soldering surface mount components.

⤷ *Mechanical skills.* You do not need any advanced mechanical skills to prepare and build the data monitor's enclosure.

⤷ *Electronics knowledge.* Although you need very little electronics knowledge to build and operate the data monitor, you should be aware of static electricity, and the damage it can cause. Observe all static precautions when handling and building the data monitor.

⤷ *Programming skills.* You do not need any programming skills to build and operate the data monitor. However, if you wish to change the data monitor's microcode or change the data monitor's host software, the data monitor's firmware is written in assembly for the Microchip PIC16C74 microcontroller. If you wish to modify the data monitor's firmware, you will need a suite of tools available from Microchip Technology Inc. Most of the PIC tools are free. This includes the assembler and simulator. Tools are available on the Microchip web page at HTTP://www.microchip.com.

The data monitor's control software is written in Borland C++ V4.51 for Microsoft Windows. If you wish to modify this software, you will need Borland's C/C++ design tools or equivalent. Be forewarned—C++ is not for the weak of heart!!

Equipment Needed

You will need, as a minimum, the following tools and equipment for the safe construction of the data monitor. Some of these elements are included with the kit and are marked as such.

- a static-free work area

- bright work light

- magnifying glass for inspection of your work

- 25-watt soldering station with medium and fine tips.

- soldering tip cleaner

- solder, fine strand

- de-soldering braid

- circuit board cleaner and cotton swabs

- tweezers

- X-acto knife, various blade styles.

- wire strippers

- small needle nose pliers

- small vise

- Dremel tool, used to cut the holes in the plastic chassis

- component lead former

The solder provided with the kit is water soluble, organic core solder. This means that you do not need any chemicals to clean the printed circuit board after assembly. Simply wash the circuit board under warm water while scrubbing with a stiff brush. An old toothbrush works fine. The top of the circuit board supplied with the kit is the side with the board name printed on it.

Note that all of the component geometries on the circuit board have one pin or pad larger than the others. This large pin identifies pin 1 of the component. Observe all static precautions when handling all of the integrated circuits. Snip all leads after soldering so that they protrude approximately 0.1 inch from the back side of the printed circuit board.

Building the Power Supply Section

Building and testing the power supply section is relatively straightforward. All components should be installed from the top side of the circuit board unless otherwise noted. The approximate construction time for the power supply section should be 45 minutes.

Here are the components you will install on the board and their functions:

- J4: 1-inch center, 5-pin right angle male connector. This connector supplies all the power connections between the circuit board and the chassis. Snap 5 pins from the 36-pin connector. Install this connector so that the pins face to the outside edge of the circuit board. This will allow the mating connector to mate with this connector.

- D1,D2: 1N4001 rectifier diodes. These diodes regulate the flow of current from the AC adapter and 9-volt battery. They insure that the AC adapter does not force current back into the 9-volt battery, and vice versa.

- U6: LM340 fixed 5-volt regulator. This integrated circuit regulates the raw input voltage down to the +5 volts needed to run the data monitor. Install this IC so that the front of the IC faces the circuit board (metal back faces up). Solder U6 into the circuit board and then bend the chip downward.

- C11: 0.22 μf capacitor. This capacitor filters the incoming voltage removing noise and transients. This capacitor is not polarized, therefore either terminal can be pin 1.

✤ C10: 0.1 µf capacitor. This capacitor filters the regulated (output) voltage removing higher frequency noise and transients. This capacitor is not polarized, therefore either terminal can be pin 1.

✤ C12: 10 µf capacitor. This capacitor filters and stabilizes the regulated (output) voltage removing lower frequency noise and transients. This capacitor is an electrolytic, therefore polarized. Pin 1, the + (positive) terminal goes into the larger pad in the circuit board.

✤ R5: 12.1 kilohm ¼-watt resistor, 1%. This resistor in conjunction with R6 forms a voltage divider. These resistors divide down the raw input voltage by a factor of (12 * 4) / (12 + 4) or 48/16 or 3. This allows a maximum input voltage of 15 volts which corresponds to a divided voltage of 5 volts. This is the maximum voltage the PIC microcontroller can handle without damage.

✤ R6: 4.02 kilohm ¼-watt resistor, 1%. This resistor in conjunction with R5 forms a voltage divider. These resistors divide down the raw input voltage by a factor of (12 * 4) / (12 + 4) or 48/16 or 3. This allows a maximum input voltage of 15 volts which corresponds to a divided voltage of 5 volts. This is the maximum voltage the PIC microcontroller can handle without damage.

Testing the Power Supply Section

Set up a bench power supply for +9.0 volts DC. Connect the positive terminal of the supply to pin 4 of J4 and the negative (ground) terminal of the supply to pin 5. Measure the output voltage of the regulator on Pin 3 of U6. This voltage should be +5.0 volts DC. Measure the voltage of VBAT_SNSE (resistor divider formed by R5 and R6). This voltage should be 3.0 volts DC.

Building the RS-232 Interface Section

As with the power supply, all components should be installed from the top side of the circuit board unless otherwise noted. Approximate construction time for this section is 25 minutes.

✎ U3: MAX232. This IC performs all the level shifting between the PC and data monitor. The RS-232 voltage levels on the cable (PC) side are typically +10 volts DC. Voltage levels on the data monitor side are +5 volts DC. Pin 1 of the IC is identified by the indent on the IC package. Pin 1 on the circuit board is identified by the larger pad.

✎ C4,5, 8, 9: 1 µf capacitors, electrolytic. These capacitors help the MAX232 chip perform the voltage level shifts needed for proper operation. These capacitors are electrolytic and therefore polarized. Pin 1, the + (positive) terminal goes into the larger pad in the circuit board.

✎ C6: 0.1 µf capacitor. This capacitor filters or decouples the +5 volt DC supply voltage removing higher frequency noise and transients. This capacitor is not polarized and either terminal can be pin 1.

✎ J1: 0.1-inch center, 9-pin right angle male connector. This connector supplies all the RS232 connections between the circuit board and the chassis connector. Snap 9 pins from the 36-pin connector. Install this connector so that the pins face to the outside edge of the circuit board. This will allow the mating connector to mate with this connector.

✎ R2: 10 ohm 1/8-watt resistor, 5%. This resistor is connected in series with the ground of the RS232 connector. Its function is a current limiting resistor. In case of an electrical fault, it insures that a large amount of current cannot flow through this connector.

Building the Microprocessor Section

All components should be installed from the top side of the circuit board unless otherwise noted. Approximate construction time is 30 minutes.

✎ U1: PIC16C74 microprocessor socket. The main processor used in the data monitor is the PIC16C74. This processor is installed in a socket to allow for easy future upgrades. Pin 1 of the socket is identified by a small arrow on the plastic. Locate the microprocessor chip and install it in the socket. Make sure you align the processor correctly with the socket and press firmly to seat it in the socket.

♨ C3: 0.1 µf capacitor. This capacitor filters or decouples the +5 volt DC supply voltage removing higher frequency noise and transients. This capacitor is not polarized and either terminal can be pin 1.

♨ R1,3, 25, 26: 10 kilohm 1/8-watt resistors, 5%. These resistors are used as pull-ups on the desired signals. These pull-ups insure that the signal "floats" to +5 volt DC when not in use.

♨ J5: 0.1-inch center, 2-pin right angle male connector. This connector supplies all the status/arm switch connections between the circuit board and the chassis connector. Snap 2 pins from the 36 pin connector. Install this connector so that the pins face to the outside edge of the circuit board. This will allow the mating connector to mate with this connector.

♨ X1: CTX6-4MHZ oscillator. This oscillator in conjunction with C1 and C2 generates the clock waveform used by the PIC. The part is not polarized, so either terminal can be pin 1. After the part is installed, solder the can to the circuit board on the large pad. This keeps the part solidly attached to the circuit board.

♨ C1, 2: 15 pF capacitor. These capacitors in conjunction with X1 generate the clock waveform used by the PIC. Since these capacitors are not polarized, either terminal can be pin 1.

♨ U2: 24LC16SN EEPROM. This EEPROM stores all data sampled by the data monitor for later analysis by the user. Pin 1 of the IC is identified by the indent on the IC package. Pin 1 on the circuit board is identified by the larger pad.

♨ C7: 0.01 µf capacitor. This capacitor filters or decouples the +5 volt DC supply voltage removing higher frequency noise and transients. Because this capacitor is not polarized, either terminal can be pin 1.

Building the LED Status Section

All components should be installed from the top side of the circuit board unless otherwise noted. Approximate construction time is 25 minutes.

- ✍ Q1,2: FMMT3904 transistors. These transistors control the flow of current through the seven-segment LEDs (U7). Solder these transistors onto the printed circuit board.

- ✍ R17, 18, 19, 20: 10 kilohm 1/8-watt resistors, 5%. These resistors limit the base current for transistors Q1 and Q2.

- ✍ R9 through 16: 91 ohm 1/8 watt resistors, 5%. These resistors limit the segment current for U7.

- ✍ U7: LN524 seven-segment LED. All status data from the data monitor is displayed through the seven-segment LED. This part is not washable with circuit board cleaner! (However water is okay.) Install the seven-segment LED display with the decimal points facing the bottom of the board, i.e., towards the PIC socket.

Installing the Personality Module Connectors

Unlike the construction steps discussed so far, all components should be installed from the bottom side of the circuit board unless otherwise noted. Approximate construction time for this section is 15 minutes.

- ✍ J3,4: 0.1-inch straight male connector. The daughter card connectors are installed from the *back* side of the printed circuit board. Snap the 8 and 10 pins from the 36-pin connector. Install J2 and J3 making sure they extend straight up from the back of the circuit board.

The printed circuit board is now completed. If you used the solder supplied with the kit, you can clean the circuit board by scrubbing it under a stream of warm water.

Now we will prepare the chassis itself.

Preparing the Chassis

Prepare the chassis for the mounting of the circuit boards and all switches and connectors. Approximate construction time for this section is three hours.

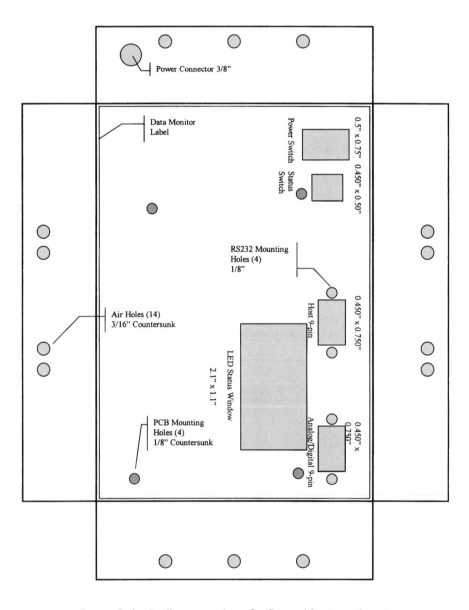

Figure B-1: Drilling template for Data Monitor chassis

The drilling template is shown in Figure B-1; a full-size file is included on the accompanying CD with this book. To prepare the template, cut along the exterior solid black line. Now stick the label carefully on the top of the data monitor's chassis, folding the "ear" of the label down over the chassis sides. The battery compartment door must be on the bottom left. This template locates the holes for the power switch, the arm/status switch, the status window, the host and input connectors, the power connector, and all air holes. Cut and drill these holes with the "weapon" of your choice, but please be careful. When the holes are completed, remove the template.

Prepare the host RS-232 connector and cable. Locate the 9-pin female socket D-connector, 9-pin 0.1-inch push-on connector, and cut 6 inches of the 9-pin ribbon cable. Separate the 9 conductors for about one inch from both ends and strip approximately 0.25-inch of insulation of each end. Tin each wire. Cut five pieces of the 1/16-inch heat shrink tubing in half and insert them over one end of the cable. Now solder the nine wires to the 9-pin connector. The 9-pin connector has pin identifiers molded into the plastic of its housing. Make sure you pay attention to the pin numbers as you solder the wires to the connector. Slide the heat shrink tubing down over the wires and solder cups and shrink them. If you don't have a heat gun, a hair dryer or match will work just fine. Now locate the 9-pin straight 0.1-inch push on connector. Cut five pieces of 1/16-inch heat shrink tubing in half and place on the other end of the ribbon cable. Solder the 0.1-inch connector on the end. Adjust and shrink the tubing. Make sure you connect pins 1 through 9 of the RS-232 connector to pins 1 through 9 of the 0.1-inch push-on connector. This completes the RS-232 host connector and cable.

Prepare the analog/digital input connector and cable. Locate the 9-pin male D-connector, the 0.1-inch push-on connector, and cut eight inches of the 9-pin ribbon cable. Separate the nine conductors for about one inch from both ends. Strip approximately 0.25-inch of insulation off each end. Tin each wire. Cut five pieces of the 1/16-inch heat shrink tubing in half and insert them over one end of the cable. Now solder the nine wires to the 9-pin connector. The 9-pin connector has pin identifiers molded into the plastic of its housing. Make sure you pay attention to the pin numbers as you solder the wires to the connector. Slide the heat shrink tubing down over the wires and solder cups and shrink them. If you don't have a heat gun, a hair dryer or match will work just fine. Now locate the 9-pin straight 0.1-inch push on connector. Cut five pieces of the 1/16-inch heat shrink tubing in half and place on the other end of the ribbon cable. Solder the 0.1-inch connector on the end. Adjust and shrink the tubing. Make sure you connect pins 1 through 9 of the input connector to pins 1 through 9 of the 0.1-inch push-on connector. This completes the input host connector and cable.

Locate the four printed circuit board ½-inch standoffs and the four pan head screws. These standoffs are used to mount the circuit board to the chassis. First, countersink the four standoff holes in the chassis so that when the screw is inserted, it is flush with top of the housing. Now mount the standoffs to the chassis.

Locate the data monitor label. To prepare the label, cut along the exterior solid black line. Also cut out the centers of all holes except for the PCB mounting holes. Now stick the label carefully on the top of the data monitor's chassis. Using an X-Acto knife, trim all of the holes in the label to match the holes you have cut into the chassis as described above. To make the label stick more securely, roughen the top of the chassis with steel wool, or fine sandpaper. Make sure you remove all dust before applying the label.

Now mount the two connector assemblies into the chassis. The RS-232 host connector should mount to the left on the data monitor and the input connector should mount to the right. Use the supplied D-sub connector hardware to mount the assemblies. Push the straight ¼-inch connectors through the holes in the chassis. Press the power switch and arm/status switch into the cover of the data monitor.

Locate the status window frame and bezel. Remove the protective plastic coating and place the bezel into the frame from the back side. Now push the two frame arms onto the frame until they snap into place. Be careful, the arms and frame are delicate. Now snap the completed assembly into the front of the data monitor.

Locate the female (plug) power connector. Cut six inches of the black and red 22-gauge wire. Solder the read and black wires to the plug as shown below. The view shown in Figure B-2 is looking into the plug connector from the plug side, not the soldercup side. Use red for the prime power pin and black for the ground pin. Cut a piece of 1/16-inch heat shrink tubing in half and place over the ends of the wire and soldercups of the power plug and shrink them. Make sure you solder the ground (black) to pin 1 and prime power (red) to pin 3. The pin numbers are molded into the plastic of the connector next to the solder cups and are the same for the plug and the mate. Mount the power plug assembly to the data monitor chassis using the hardware supplied.

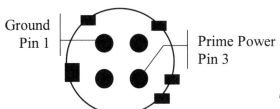

Ground Pin 1

Prime Power Pin 3

Figure B-2: Pin-out of AC adapter

Cut two 4-inch pieces of the black 22-gauge wire and solder one to each terminal of the power switch. Leave the other ends of the wires dangling for now.

Cut two 8-inch pieces of the red 22-gauge wire and solder one to each terminal of the arm/status switch. Slide half pieces of 1/16-inch heat shrink tubing on the other ends of the wires. Attach a 2-pin 0.1-inch push-on connector to the ends of the wires. The order of the wires does not matter. Position the heat-shrink tubing over the solder cups and wires and shrink.

Remove the battery compartment door from the bottom of the data monitor and insert the 9-volt battery connector through one of the holes in the chassis. Close the cover. Cut the length of the wires to five inches.

Now attach the 0.1-inch 5-pin push-on connector to the end of the power connector leads. Snap 5 pins from the 36-pin connector. Make sure you push a half piece of 1/16-inch of heat shrink tubing on each wire before you solder it to the connector. Attach the red wire from the power connector to pin 1 of the 5-pin connector. Attach the red power lead from the 9-volt battery to pin 2 of the 5-pin connector. Attach one side of the power switch wires to pin 3 of the 5-pin connector. Attach the other side of the power switch wires to pin 4 of the 5-pin connector. Finally, attach both black wires from the power plug and 9-volt battery to pin 5 of the 5-pin connector. It will be easier if you attach one wire to the connector and then attach the second to the first. For this connection use the 1/8-inch heat-shrink tubing. Push the heat-shrink tubing down over the rest of the pins and shrink.

Locate the AC adapter that came with the data monitor. If there is a plug on the end of the wire on the output of the adapter, cut it off. Attach the male power plug to the end of this cable. Refer to the AC adapter pin-out for the connections. Make sure you solder the ground (black) to pin 1 and prime power (black with white stripe) to pin 3. The pin numbers are molded into the plastic of the connector next to the solder cups. Double check the output of the AC adapter with a voltmeter before soldering. Unscrew the plastic housing from the connector to expose the solder cups. Strip the power adapter wires about ¼-inch. Place half pieces of 1/16-inch heat-shrink tubing over the wires. Solder the wires to the connector as shown, making sure you place the plastic housing cover on the wire before you solder. Push the heat-shrink tubing down over the pins and shrink. Replace the plastic cover.

This completes the chassis and cable preparation.

Mounting the Circuit Board into the Chassis

Now that the cable assemblies and circuit board are complete, we are ready to mount the circuit board into the chassis. Take the cover of the data monitor and turn it over, exposing the back of the switches. Approximate construction time for this section is 10 minutes.

✍ Attach the 5-pin power cable to the circuit board (J4). Make sure you keep pin 1 of the cable with pin 1 of the connector.

✍ Attach the 9-pin RS-232 cable to the circuit board (J1). Make sure you keep pin 1 of the cable with pin 1 of the connector.

✍ Attach the 2-pin arm/status cable to the circuit board (J5). Make sure you keep pin 1 of the cable with pin 1 of the connector.

Now turn the circuit board over and place inside the chassis on top of the standoffs. Using the screws provided, mount the circuit board to the chassis.

Congratulations! You have completed assembling the data monitor motherboard and chassis. Even though the data monitor cannot do much yet without a personality module installed, a small test can be performed to make sure that the data monitor is functional.

Place the back cover on the data monitor and attach it with a few screws. Connect the AC adapter power plug or connect a 9-volt battery to the data monitor. Turn on the power switch and press the arm/status button. You should see two horizontal lines on the data monitor's status window. This tells us that the data monitor has passed all its self tests and is ready for communications with the host. Install the data monitor software and start it up. Choose Download -> Voltage. You should get a pop-up window specifying a system voltage between 10 and 12 volts.

If you don't receive the System Voltage pop-up, check your work. Make sure all components which are polarized are installed correctly. Check all solder joints and connections.

General I/O-Temperature Personality Module Assembly Instructions

Locate the General I/O-Temperature personality module printed circuit board. The top of the circuit board is the side with the board name printed on it. Note that all of the component geometries on the circuit board have one pin or pad larger than the others. This large pin identifies pin 1 of the component. All components should be installed from the top side of the circuit board unless otherwise noted. Observe all static precautions when handling all of the integrated circuits.

Building the Analog Input Section

The approximate construction time for this section is 40 minutes.

✋ U2, 3: LM6492B operational amplifier. This integrated circuit buffers all analog inputs from the external 9-pin D connector to the data monitor. The gain of the op amp, input attenuation, and filtering can be set though the use of the support resistors and capacitors. Solder these parts to the circuit board noting the position of pin 1. Pin 1 is identified by an indent on the package. Pin 1 on the circuit board is identified by the larger signal pad.

✋ R5 through R20: 10 kilohm, 1/8-watt resistors, 5%. Resistors R7, 8, 11, 12, 15, 16, 19, and 20 set the gain produced by the op amp. R5, 9, 13, and 17 form part of the filter for the analog inputs. R6, 10, 14, and 18 have two purposes; they are used as pull-downs insuring that the analog input does not "float" (produce erroneous results) when no external connection is made and they also form a voltage divider allowing the user to attenuate the input voltage range that the op amp sees.

✋ C3 to C6: 0.1 µf capacitors. These capacitors filter the analog inputs in conjunction with resistors R5, 9, 13, and 17. These capacitors are not polarized so either terminal can be pin 1.

✋ C7, 14: 0.1 µf capacitors. These capacitors filter or decouple the +5 volt DC supply voltage removing higher frequency noise and transients. These capacitors are not polarized and either terminal can be pin 1.

Building the Digital Input Section

The approximate construction time for this section is 15 minutes.

✋ U4: D74HC14 Schmitt trigger inverter. This integrated circuit buffers all the digital inputs from the external 9-pin D connector to the data monitor. The input attenuation plus some filtering is performed with the use of support resistors and capacitors. Solder this part to the circuit board noting the position of pin 1. Pin 1 of U4 is identified by an indent on the package. Pin 1 on the circuit board is identified by the larger signal pad.

✎ R21 to 24: 1 kilohm 1/8-watt resistors, 5%. These resistors in conjunction with resistors R25 through 28 form a voltage divider which can attenuate the digital input. They also form a low-pass filter in conjunction with capacitors C9 to C12.

✎ R25 through 28: 10 kilohm 1/8-watt resistors, 5%. These resistors in conjunction with resistors R21 to 24 form a voltage divider which can attenuate the digital input. They also form a low-pass filter in conjunction with capacitors C9 to C12.

✎ C9 to C12: 0.1 µf capacitors. These capacitors filter the analog inputs in conjunction with resistors R21, 22, 23, and 24. These capacitors are not polarized and either terminal can be pin 1.

✎ C8: 0.1 µf capacitors. This capacitor filters or decouples the +5 volt DC supply voltage removing higher frequency noise and transients. This capacitor is not polarized and either terminal can be pin 1.

Building the Temperature Sensor Section

The approximate construction time for the temperature sensor section is 30 minutes.

✎ Q7: LM50 temperature sensor. This integrated circuit reacts to the current ambient temperature and produces a linear output voltage in proportion to that temperature. Solder this part to the circuit board where indicated.

✎ U1: LT1006 op amp. This op amp is used to amplify the output from the LM50 temperature sensor. Its gain is set to 2.5 through the resistors R3 and R4. Solder this part to the circuit board noting the position of pin 1. Pin 1 of U1 is identified by an indent on the package. Pin 1 on the circuit board is identified by the larger signal pad.

✎ R3: 1 kilohm 1/8-watt resistor, 5%. This resistor in conjunction with R4 sets the gain of op amp U1.

↳ R4: 1.5 kilohm 1/8-watt resistor, 5%. This resistor in conjunction with R3 sets the gain of op amp U1.

↳ R2: 10 kilohm 1/8-watt resistor, 5%. This resistor limits the current into the + terminal of op amp U1. It also tries to match the input impedance between the + and - terminals.

↳ C1: 0.1 µf capacitor. This capacitor removes any noise or transients from the output of the LM50 temperature sensor. This capacitor is not polarized and either terminal can be pin 1.

↳ C13: 0.1 µf capacitor. This capacitor filters or decouples the +5 volt DC supply voltage removing higher frequency noise and transients. This capacitor is not polarized and either terminal can be pin 1.

↳ J4: 3 pin 0.1-inch, 0.025-inch square right angle male connector. This connector is used to connect either the analog 1 input or the temperature sensor to the first analog input of the data monitor. Placing a jumper or shorting pad across pins 1 and 2 connect the external analog input to the data monitor. Placing a jumper across pins 2 and 3, connect the temperature sensor to the data monitor. Snap 3 pins from the 36-pin connector. Install this connector with the pins facing towards the side of the printed circuit board. Now locate the 2-pin jumper and push onto pins 2 and 3 of the 3-pin header. This will connect the temperature sensor into analog 1 of the data monitor.

Attaching the Host Connectors

The approximate construction time for this section is 20 minutes.

↳ J2: 8-pin straight 0.1-inch female connector. Install this connector from the component side of the printed circuit board. The pins should face up. Make sure that the pins are at a right angle from the circuit board. Snap 8 pins from the 36-pin connector.

↳ J1: 10-pin straight 0.1-inch female connector. Install this connector from the component side of the printed circuit board. The pins should face up. Make sure that the pins are at a right angle from the circuit board. Snap 10 pins from the 36-pin connector.

   ✍ J3: 9-pin right angle 0.1-inch male connector. Install this connector from the component side of the printed circuit board. The pins should face towards the side of the circuit board. Snap 9 pins from the 36-pin connector.

   ✍ R1: 10 kilohm 1/8-watt resistor, 5%. This resistor is used to set the ID of the General I/O-Personality module. The ID of this module is set to 0b001.

The printed circuit board is now completed. If you used the solder supplied with the kit, you can clean the circuit board by scrubbing it under a stream of warm water.

Congratulations again! You have now completed the General I/O Temperature Module. Now we need to complete all cable assemblies and install the Personality module into the data monitor.

Mounting the General I/O-Temperature Module into the Data Monitor

The approximate construction time for this section is 5 minutes.

   ✍ Attach the 9-pin external cable to the circuit board (J1). Make sure you keep pin 1 of the cable with pin 1 of the connector.

   ✍ Now push the Personality module directly into the mating connectors on the data monitor. This completes the installation of the data monitor.

   ✍ Locate the foam square. Attach this foam to the inside of the back cover of the chassis using glue or double-sided tape. This foam insures that the Personality module circuit board is held tightly in the closed chassis.

   ✍ Replace the bottom cover of the chassis and tighten all screws.

Assembling the Input Breakout Connector

The approximate construction time for this section is 10 minutes.

↬ Locate the five colored test connectors. Cut each connector exactly in half to create ten test connectors. Place one of the black leads aside; this one is extra.

↬ Locate the test connector labels. Place the AN1, 2, 3, and 4 labels on a green, red, white, and yellow connector, respectively. Place the DIG1, 2, 3, and 4 label on a green, red, white, and yellow connector, respectively. Place the GND label on the black test connector.

↬ Strip each test connector lead approximately ¼-inch and tin. Slide approximately ½-inch of heat-shrink tubing over each wire.

↬ Locate the 9-pin D female (socket) connector. Solder the test connector leads to pins 1 through 9 of the D connector solder cups in the following order: DIG1, 2, 3, 4, GND, AN1, 2, 3, 4 respectively. Slide half pieces of the 1/16-inch heat shrink tubing down over each solder cup and shrink.

↬ Locate the 9-pin D female plastic housing. Insert the test connector assembly into the housing and secure the wires with the retainer clip. Close the connector housing by snapping the shell closed. Add the retainer clips to the 9-pin D connector.

The test connector assembly is now complete. Use this connector assembly to connect the data monitor to the signals you wish to monitor.

Testing the Data Monitor

Now that you have fully completed assembling the data monitor, we must test it to make sure it is fully functional.

↬ Open the battery compartment door and install a fresh 9-volt battery. You can use an alkaline or NiCad battery. Close the compartment door.

✍ Turn on the power to the data monitor. Press the Arm/Status button and insure that the display shows two single bars. These lines indicate that the data monitor has successfully passed all its power-on tests and is ready for communications to and from the host computer. Turn the data monitor off.

✍ Now plug the AC adapter into the wall and connect the plug to the data monitor.

✍ Turn on the power to the data monitor. Press the Arm/Status button and insure that the display shows two single bars. These lines indicate that the data monitor has successfully passed all its power-on tests and is ready for communications to and from the host computer. You can leave the 9-volt battery installed when running from the AC adapter. The data monitor contains protective circuitry to isolate the battery from the AC adapter source.

✍ Connect the host 9-pin cable from the PC COMPORT (either COM1 or COM2) to the host connector on the data monitor.

✍ Install the data monitor software as described in the data monitor User's Guide and Technical Reference Manual. When installed, double click on the data monitor icon to run the data monitor software.

✍ Select the Configure -> Comport pull-down menu pick , and select the active COMPORT you wish to use and click OK.

✍ Select the Configure -> Setup pull-down menu pick, and configure the data monitor to record all of the digital and analog inputs. You can leave all the other options at their default. The data monitor will default to store immediate with 30 milliseconds between samples. The Setup screen should look like the screen shown in Figure B-3.

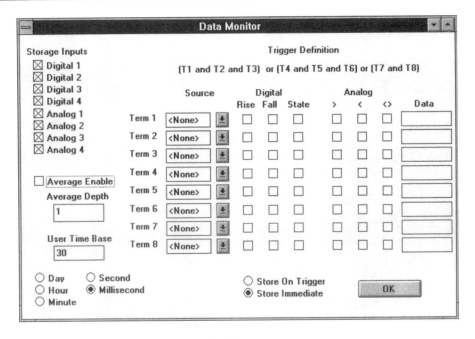

Figure B-3: Test setup screen

✎ Select the Configure -> Initialize pull-down menu pick to initialize the data monitor with the selected configuration. You should receive a data monitor initialized pop-up if everything is okay. If not, check your cables and connections, internal and external. Make sure you haven't plugged some of the internal cables on backwards.

✎ Now press and hold the Arm/Status button. You should see two three-bar characters show up in the status window of the data monitor. Release the button. This press of the Arm/Status button "arms," or begins, the store cycle in the data monitor. The complete store cycle should take less than a minute. Let the data monitor store until it is full. You can monitor the progress of the store cycle by pressing the Arm/Status button. When pressed, the status window will display the EEPROM percentage full. When the status window shows "00.", the data monitor EEPROM is full and ready for download to the host. Note that you do not need to wait for the EEPROM to fill completely before downloading; you can download at any time.

✎ Select the Download -> Download pull-down menu pick. A pop-up menu will appear. Type the name of the file you wish to save the data in. TEST.DNL will suffice for now. Also select a destination directory. Press the OK button and DMON.EXE will download the data monitor and place the data in the text file named above.

✎ Now check the contents of the download file. Because we have left all of the data monitor's inputs unconnected, all digital and analog inputs should report "0" except for AN1, which should report the current temperature. This value should be around 85 to 95, depending on the ambient temperature. This represents a temperature of approximately $((88/255) * 5.0$ volts$) / 2.5 - 0.5$ volts, or 21 degrees Celsius. Refer to the data monitor's Users Guide and Technical Reference Manual for more details on the above calculation. The contents of the download file should look like the following:

```
Data Monitor Download Data (V1.0) — JV Enterprises

Storage Mode  :: Immediate
Trigger Mode  :: None
Average Off   :: —
Time Stamp    :: 20:50:22.070
Date Stamp    :: 01\02\97
User Time Scale :: 30 milliseconds
```

Time	Dg1	Dg2	Dg3	Dg4	An1	An2	An3	An4
20:50:25.297	0	0	0	0	88	0	0	0
20:50:25.327	0	0	0	0	88	0	0	0
20:50:25.357	0	0	0	0	88	0	0	0
20:50:25.387	0	0	0	0	88	0	0	0
20:50:25.417	0	0	0	0	88	0	0	0

(file continues)

✍ Now attach the Input Breakout Connector to the data monitor. Attach a +4.5 volt bench supply to the DIG1 input and the ground to GND.

✍ Select the Configure -> Setup pull-down menu pick, and configure the data monitor to record all of the digital and analog inputs as before.

✍ Select the Configure -> Initialize pull-down menu pick to initialize the data monitor with the selected configuration. You should receive a data monitor initialized pop-up if everything is okay.

✍ Now press and hold the Arm/Status button. You should see two three-bar characters show up in the Status window of the data monitor. Release the button. This press of the Arm/Status button "arms," or begins, the store cycle in the data monitor. The complete store cycle should take less than a minute. Let the data monitor store until it is full. You can monitor the progress of the store cycle by pressing the Arm/Status button. When pressed, the status window will display the EEPROM percent full. When the status window shows "00.", the data monitor EEPROM is full and ready for download to the host. You do not need to wait for the EEPROM to fill completely before downloading. You can download at any time.

✍ Select the Download -> Download pull-down menu pick. A pop-up menu will appear. Type the name of the file you wish to save the data in. TEST.DNL will suffice for now. Also select a destination directory. Press the OK button and DMON.EXE will download the data monitor and place the data in the text file named above.

✍ Now check the contents of the download file. Now, all digital and analog inputs should report "0" except for AN1 which should report the current temperature and DIG1 which should report "1" instead of "0". The contents of the download file should look like the following:

```
Data Monitor Download Data (V1.0) — JV Enterprises

Storage Mode :: Immediate
Trigger Mode :: None
```

```
Average Off   :: —
Time Stamp    :: 20:50:22.070
Date Stamp    :: 01\02\97
User Time Scale :: 30 MilliSeconds
```

Time	Dg1	Dg2	Dg3	Dg4	An1	An2	An3	An4
20:50:25.297	1	0	0	0	88	0	0	0
20:50:25.327	1	0	0	0	88	0	0	0
20:50:25.357	1	0	0	0	88	0	0	0
20:50:25.387	1	0	0	0	88	0	0	0
20:50:25.417	1	0	0	0	88	0	0	0

```
(file continues)
```

✎ Repeat the above six steps three more times for each of the other digital inputs (D2, 3, 4). Verify that the data monitor reports the correct logic state for each of the tests. At this point all of the digital inputs are verified and operational. If any test failed, check all the connections and correct any mistakes.

✎ Now attach a +2.5-volt bench supply to the AN2 input and the ground to GND.

✎ Configure and initialize the data monitor as before. Arm and download the data monitor as before. AN2's reported value should be approximately (2.5v/5.0v) * 255, or 128 (+-1). Refer to the data monitor's Users Guide and Technical Reference Manual for more details on the above calculation. The download file should look like the following.

```
Data Monitor Download Data (V1.0) — JV Enterprises

Storage Mode  ::  Immediate
Trigger Mode  ::  None
Average Off   ::  —
Time Stamp    ::  20:50:22.070
Date Stamp    ::  01\02\97
User Time Scale ::  30 MilliSeconds

Time              Dg1   Dg2   Dg3   Dg4   An1   An2   An3   An4

20:50:25.297      0     0     0     0     88    128   0     0

20:50:25.327      0     0     0     0     88    128   0     0

20:50:25.357      0     0     0     0     88    128   0     0

20:50:25.387      0     0     0     0     88    128   0     0

20:50:25.417      0     0     0     0     88    128   0     0

(file continues)
```

✍ Repeat the above two steps for the other two analog inputs AN3 and 4.

✍ For completeness' sake, you can swap the jumper on the General I/O-Temperature Module to pins 1 and 2, connect the 2.5-volt power supply to AN1, and run a store and download cycle as described above.

At this point the data monitor is fully operational and ready for use. Congratulations!

APPENDIX C
Data Monitor Temperature Sensor Application Note

This application note details the operation of the built-in temperature sensor of the data monitor. This sensor allows you to directly monitor ambient temperature within a range of -10 degrees Celsius to 65 degrees Celsius (14 degrees Fahrenheit to 150 degrees Fahrenheit). The holes in the sides of the data monitor allow the air to circulate around the temperature sensor allowing the sensor, to track temperature variations as quickly as possible.

In order to verify the operation of the data monitor, as well as become familiar with its operation, this note will walk you through the steps of setting up, configuring, monitoring, downloading, and analyzing the data monitor's output. We will monitor the temperature of an open room, the inside of a refrigerator, and then the same open room again.

The application note is based on the following assumptions.

1. You have successfully completed the construction and testing of the data monitor and General I/O-Temperature personality module kit described in Appendices A and B.

2. You have completely read Appendices A and B.

3. Your data monitor has the General I/O-Temperature personality module installed.

4. The jumper on the General I/O-Temperature personality module is set to select the temperature input instead of Analog 1.

You can check to make sure which personality module is installed in the data monitor by removing the back cover of the data monitor, removing the personality module, and reading the label on the personality module. While you have the personality module out, check the position of the jumper. The jumper should be installed on pins 2-3 as shown in Figure C-1.

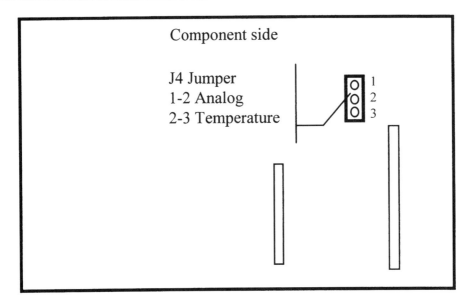

Figure C-1: Personality module jumper location

Before we get into the details of setting up and configuring the data monitor, let's talk a little about the temperature sensor itself. The temperature sensor installed on the Personality module is a National Semiconductor LM50. This sensor is capable of detecting temperatures from -10 to +65 degrees Celsius. This temperature sensor exhibits a linear output over the full range of temperatures. The output of the LM50 is represented by the following formula:

Output Voltage = 0.5volts + (10mv)*(Degrees Celsius)

For example, a temperature of 20 degrees Celsius would yield an output voltage of .5 + (.01*20), or .7 volts.

The output of the temperature sensor is then passed through a non-inverting op amp with a fixed gain of 2.5. This expands the output of the sensor to cover the full range of the data monitor's A/D converter. (Remember that the reference voltage for the A/D converter inside the data monitor is 5.0 volts.) That means that the digital value reported by the data monitor is represented by the following formula:

*Output value = (Input voltage / 5.0v) * 255*

Continuing our example, a temperature of 20 degrees Celsius produces an output of .7 * 2.5 = 1.75 volts. This would be reported by the data monitor as an A/D value of 89 out of 255.

With the op amp circuit configured as shown, the smallest temperature the data monitor can distinguish is 0.787 degrees Celsius, or 1.42 degrees Fahrenheit. If we desired more precision, we would have to change the op amp circuit to focus on the temperature range we desired. Keep in mind this would reduce the overall range of the circuit. We must trade off total range for precision.

In its current configuration, the temperature sensor is not meant to monitor temperatures that change more than 0.3 degrees Celsius per minute. This makes it ideal for monitoring ambient temperatures and environments where the temperature does not vary too quickly. The limitation on degrees per minute is due to the small holes in the chassis of the data monitor. If you desire a temperature sensor that can react faster than 0.3 degrees per minute, you would need an external sensor that has a very small mass, or provide more air flow over the existing sensor.

Setting Up the Data Monitor

The first step in setting up the data monitor is to load the data monitor software into a PC. Your PC can be running Windows 3.1x or Windows 95. Insert the installation disk into drive A, and run A:\setup. Follow the directions on the screen to install the software.

Now locate the following items:

↳ the data monitor

↳ the host serial cable (9 pin)

↳ a fresh 9-volt battery

Remove the battery compartment door and install the 9-volt battery and re-place the door. A fresh 9-volt battery should supply enough power to run the data monitor for approximately 9 hours. Now turn on the data monitor's prime power switch and press the Arm/Status button. You should see two dashes displayed in the status window. These tell us that the data monitor is alive and well and ready to accept commands from the software.

Connect the serial cable from the data monitor to COM1 or COM2 of your PC. The data monitor only supports COM1 or COM2. This completes the setup of the data monitor hardware.

Configuring the Data Monitor

Run the data monitor's configuration software by double clicking on the data monitor icon. Figure C-2 shows the initial arrangement of the data monitor Initial Setup screen.

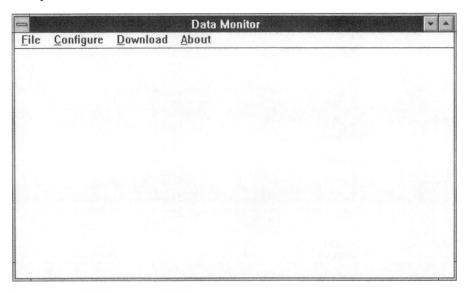

Figure C-2: Data Monitor Initial Setup screen

Now we will configure the data monitor to sample and store the temperature sensor input. Refer to Figure C-3 during the following discussion.

First, we want the data monitor to sample the temperature sensor. This temperature input has been set to appear on Analog 1 by the use of the jumper on the General I/O - Temperature personality module. Therefore, select Analog 1 for the Storage Inputs.

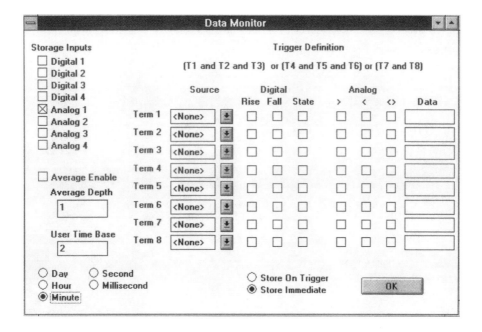

Figure C-3: Data Monitor Setup screen

For this particular example we do not need averaging enabled, so do not check the Average Enable box. The Average Depth value is ignored if the Enable is not checked.

Select a Time Base of two minutes.

For this particular example we will not select a trigger. We will sample the temperature every 2 minutes. Therefore, select Store Immediate.

Your configuration screen should look as shown in Figure C-3.

Click OK to accept the configuration. Now select the pulldown menu Setup -> Initialize. This will initialize the data monitor and prepare it for monitoring.

Monitoring

Now that the data monitor is configured, we are ready to arm the unit to begin the sample and store process. Place the data monitor on the table. Press and hold the Arm / Status button. You should see two three-bar characters in the status window. These characters tell you that the data monitor has been armed, and the sample and store process has begun. Subsequent presses of the Arm / Status button will show you the percent of EEPROM (memory) that has been filled. The percents reported range from 00 to 99. When the data monitor is full, the status window will display 00. Note the " . " on the right hand side. Let the data monitor sit on the table for 20 minutes, then place the unit in a refrigerator for another 20 minutes, and finally back on the table for 20 minutes more. This should fill the data monitor's EEPROM approximately 1%. With this configuration, the data monitor will be able to store data for approximately 2029 samples, or 67 hours and 38 minutes. The data monitor allows you to download the collected data at any time during the sample process. However, it will not be able monitor and store new information during the download.

Downloading

Now that we have filled the EEPROM with some data, let's download the information so that we can review it. Downloads are used to transfer recorded information from the data monitor to the host for analysis.

Downloaded information is recorded in a .DNL file specified by the user. It is an ASCII text file and can be pulled into any standard editor or word processor. It is recommended that the user import the download file into a spreadsheet such as Microsoft Excel® and use its superior graphics capabilities to plot or analyze the results. This is what we will do.

Connect the data monitor back up to the PC using the serial cable and double click the Data Monitor Application. Now select the Download -> Download item from the pulldown menu and enter the file name you wish to save the data as. When you are ready, hit OK. The download is complete when the hourglass disappears. If there is trouble talking to the data monitor, the application will let you know.

Your download file should look something like the following (only the first few samples are listed):

```
Data Monitor Download Data (V1.0) — JV Enterprises

Storage Mode      ::  Immediate
Trigger Mode      ::  None
Average Off       ::  —
Time Stamp        ::  21:15:28.290
Date Stamp        ::  05\25\97
User Time Scale   ::  2 Minutes

Day           Time                  An1
00            21:17:33.192          90
00            21:19:33.192          90
00            21:21:33.192          90
00            21:23:33.192          90
00            21:25:33.192          90
00            21:27:33.192          90
00            21:29:33.192          90
00            21:31:33.192          90
00            21:33:33.192          90
00            21:35:33.192          90
00            21:37:33.192          90
00            21:39:33.192          90
00            21:41:33.192          90
00            21:43:33.192          88
00            21:45:33.192          86
00            21:47:33.192          85
00            21:49:33.192          83
00            21:51:33.192          82
00            21:53:33.192          81
00            21:55:33.192          80
00            21:57:33.192          79
00            21:59:33.192          78
...
```

Analyzing

When the download is complete, start your spreadsheet program and import the data file that you specified above. This file is specially formatted just for this purpose. All data is stored as tab delimited data.

Once we have the data imported into the spreadsheet, we must turn the A/D samples recorded into actual temperatures. To do this, we must use the following formula

Temperature = ((((A/D Value / 255) * 5.0 volts) / 2.5) - .5 volts) * 100

This formula takes the A/D value from the data monitor and converts it to a voltage based on the microprocessor's internal 5.0-volt reference. This is done by dividing the A/D value by 255 and multiplying the result by 5.0-volts. This result is then divided by 2.5 to remove the gain added by the op amp. Next remove the 0.5-volt offset added by the LM50. Then multiply by 100 since the LM50 temperature sensors output is reported in steps of 10 mv/C.

Do this for all of the sample points imported and plot a graph of the results. You should achieve results similar to the ones shown in Figure C-4.

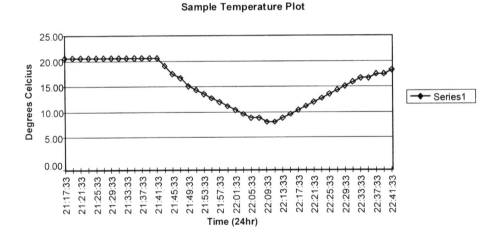

Figure C-4: Data Monitor trigger example

Variations

Now let's re-run the previous test with different trigger parameters. Let's say we want to store the temperature every time it changes more than 5 degrees Celsius in any direction.

Now we have to convert the 5-degree limit we have set to A/D bits. Previously we had determined that 1 LSB corresponds to .787 degrees Celsius. Therefore, 5 degrees corresponds to 6 LSBs (it's really 4.7 degrees Celsius, but that's close enough).

We want to select the Store on Trigger button here. This will cause the data monitor to store data only when the trigger conditions are satisfied. Note that all triggers that we don't use, i.e. those set to <none>, fall out of the trigger equation. If we had selected the Store Immediate button with a trigger term defined, the data monitor would wait for the trigger condition to be satisfied, and then store data based solely on the User Time Base. We have left the User Time Base at its default value since it is not used by the data monitor as configured. The setup screen should look as shown in Figure C-5:

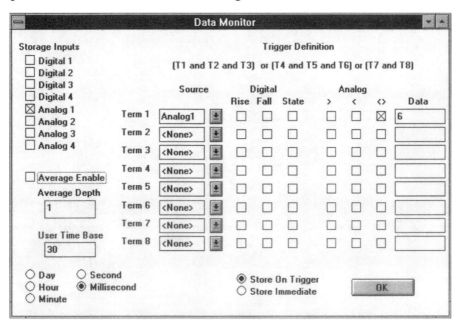

Figure C-5: Data Monitor Setup screen

Initialize, and arm, and download the data monitor as before. Note that the data monitor's EEPROM will be a lot less full at the end of this test than the last. This is due to the fact that the data monitor only has to store data when the trigger is satisfied, not at a fixed interval as before.

```
Data Monitor Download Data (V1.0) — JV Enterprises

Storage Mode    :: Trigger
Trigger Mode    :: Calculated
Average Off     :: —
Time Stamp      :: 13:33:51.290
Date Stamp      :: 05\26\97
User Time Scale :: 30 MilliSeconds

Day         Time                An1
00          13:33:52.856        88
00          14:38:42.345        82
00          14:50:07.867        76
00          15:04:49.469        82
00          15:34:48.419        88
```

The analyzed data should look something like the graph in Figure C-6.

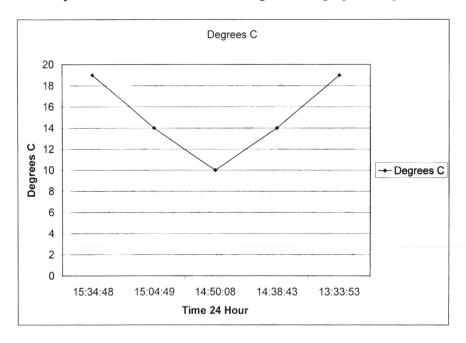

Figure C-6: Data Monitor trigger example #2

Glossary

Acetone: Acetone is a cleaner used to remove the resist layer after etching a circuit board as well as a good general purpose cleaner. It does not leave behind a residue. It is available at hardware stores. Use by applying on a cotton cloth and rubbing the board to be cleaned. Make sure you have proper ventilation when using. Acetone is flammable.

Ammonium Persulphate: Ammonium persulphate is used to remove or "etch" copper from a printed circuit board. It is supplied in dry form, and mixed with water when ready to use. A mixed solution has a shelf life of approximately six months.

Annular Ring: The circle of copper around the hole of the through-hole component minus the drill hole size.

Anvil: An anvil is used when installing barrels (our attempt at a via) in printed circuit boards. The anvil holds the barrel steady while it is formed from the other side.

Aperture: An aperture is a small window that controls the passing of light similar to the one in a camera. Apertures are used in the generation of photoplots.

Artwork: An artwork is a 1:1 reproduction of the traces, pads and through holes used to create a printed circuit board. It can be created in a multitude of ways such as CAD programs or just drawing on a piece of paper.

Barrel (Eyelet): A barrel or eyelet is used to make a connection from a trace on one side of a printed circuit board to the other. The barrel is placed through the circuit board and swaged or formed with a form tool.

BOM: Bill of Materials. This is a list of all the components (electrical and non-electrical) contained in a design. It is usually automatically generated from a schematic capture tool.

CAD: Computer Aided Design.

Component Name: Also referred to as a *symbol name*. Every electronic symbol you generate needs to have a unique name that the schematic capture tool references. The component name is similar to a file name for a file on a PC.

Component Side: This is the side of the printed circuit board that the components are normally placed on or through.

Contact Frame: A contact frame is a device that holds a circuit board and artwork(s) steady while the circuit board resist layer is exposed. It consists of a hard backboard, a piece of cardboard, a piece of glass, and two clamping bars to hold it all together.

Control Section: Control sections are used to monitor the development of a resist layer. They are nothing more than a section of a printed circuit board that is not covered by an artwork (in our case paper) when exposed.

Diffusion, Light: Light diffusion is a process where light scatters around an object, striking an object immediately behind it. Light diffusion during the exposing of the resist layer will cause fuzzy and thin traces and pads.

Electronic Symbol: An electronic symbol is a drawing of a real-world component. The format of the drawing conforms to the requirements of the schematic capture tool.

Electroplating: Electroplating is an electro-chemical process that involves the deposit of a metal onto another metal. In the case of printed circuit boards, it usually refers to a tin-based alloy that is deposited on copper.

Etchant: Etchant is the chemical that is used to remove unwanted copper from a printed circuit board.

Etching: Etching is the process of removing unwanted copper from a printed circuit board (and hopefully leaving the copper you do want!).

Exposure Cone: An exposure cone is a device that helps control and direct the light from an exposure lamp to an exposure frame. Plans for an exposure cone are given in the plans section.

Exposure Frame: An exposure frame is a device which holds a circuit board and artwork(s) steady while the circuit board resist layer is exposed. It consists of a hard backboard, a piece of cardboard, a piece of glass, and two clamping bars to hold it all together.

Eyelet (Barrel): A barrel or eyelet is used to form a connection from a trace on one side of a printed circuit board to the other. The barrel is placed through the circuit board and swaged or formed with a form tool.

Ferric Chloride: Ferric chloride is used to remove or "etch" copper from a printed circuit board. It is supplied in dry or mixed form. It stains most metals, including stainless steel, as well as skin.

Flash: Passing light through a photoplotter aperture without moving the aperture is known as a "flash."

Footprint: A footprint is a physical representation of a component. Footprints are used in a layout tool.

Form Tool: A form tool is used in conjunction with an anvil and eyelets (barrels). It is used to form the end of the eyelet into various shapes.

FR-4: FR-4 describes the material most often used as the substrate in printed circuit boards. It is a fiberglass resin based material that is very hard and resistant to temperature and chemicals.

Gerber File: The standard file format used to transfer printed circuit board design data to a fabrication house. Gerbers files can be standard or extended.

Goof-off: Goof-off is a commercial cleaner that can be used to clean the resist layer off a circuit board after it has been etched. It is also a good general-purpose cleaner.

Hot Air Leveling: Hot air leveling is a process that is used to level the solder or tin plate on a printed circuit board. It is achieved by dipping a circuit board in molten solder and than blowing a stream of hot air over it to remove the excess solder. Don't try this at home!

Layer: The definition of a layer is very important in the creation of footprints and padstacks. Padstacks are made up of obstacles that are placed on specific layers. In this way the user can control what connections are made on a specific layer of a circuit board.

Layout: Circuit board layout is the process of placing the footprints of all the components in your design on a circuit board, usually in some logical manner, and connecting them with traces.

MSDS: Material safety data sheet. Read it!

Negative: A negative refers to the type of artwork used in a circuit board fabrication process. A negative is opaque (or black) where we do not want copper, and clear (or white) where we do want copper.

Netlist: A netlist is usually an ASCII (text) document defining the connections between symbols in your schematic.

Obstacle: An outline or shape that represents an object on a circuit board that must be taken into account during routing or placement.

Oxidation: Oxidation is a process where a surface oxidizes or corrodes when exposed to the air.

Package Name: The package name on an electronic symbol describes the component's package type or footprint (for example, DIP16 or PLCC44).

Padstack: A padstack is a physical representation of a component pin. A padstack is made up of one or more obstacles placed on different layers. For example, a padstack could contain obstacles for the top layer of the circuit board, the inner layers, solder mask and paste mask layer.

Pad: A pad is a piece of copper, usually round or rectangular, where the leads of components are soldered. You can have surface mount pads or through-hole pads. Surface mount pads are not drilled, while through-hole pads are.

Paste Mask: Solder paste must be applied to all surface mount component pads before the board is stuffed and sent through a solder reflow or infrared oven. This mask layer tells the fabrication house exactly where the paste needs to go.

Photoplot: A photoplot is a commercially produced artwork. It is nothing more than a high quality view-graph or transparency.

Photoplotter: A plotter that makes photoplots. A photoplotter can be a vector type (old and not used much any more) or raster (like your computer printer).

Pin: A pin on an electronic symbol is equivalent to a pin on a real device. Pins are where the inputs and outputs of a device are connected to the rest of the circuit.

Positive: A positive refers to the type of artwork used in a circuit board fabrication process. A positive is opaque (or black) where we want copper, and clear (or white) where we do not want copper.

Presensitized: Presensitized circuit boards refer to the resist layer on the board. A presensitized circuit board is coated with a resist layer when you buy it. It is very sensitive to light and should be stored in a cool, low light area. Circuit boards can be presensitized for positive or negative artworks.

Reference Designator: Layout tools reference each device in a design by the symbols reference designator (also known as a "Ref Des"). Each component in a design must have a unique reference designator.

Registration: Registration refers to the alignment of the top and bottom artwork layers of a printed circuit board. Layers must be registered properly before exposing and etching can occur. Proper registration insures that all holes (or vias) will line up when drilled.

Resist Developer: Resist developer is used to develop the resist layer after exposing. When developed, a positive resist layer will harden where light does not strike it, and will dissolve where light strikes it.

Resist Layer: A resist layer allows us to control the removal of copper during the etching process. Circuit boards described in this manual are presensitized with a positive resist layer.

Resistivity: The longitudinal electrical resistance of a uniform rod of unit length and unit cross-sectional area; the reciprocal of conductivity. (What a mouthful that was!)

Router: (a) A high-speed (20,000 RPM) machinist's tool used to remove large and small amounts of hard material. This can be a very dangerous tool if not used properly.

(b) A piece of software used to connect all the placed component footprints in your design with copper traces or wires.

Routing: Circuit board routing is the process of connecting all the placed component footprints in your design with copper traces, or wires. This step usually occurs after the layout phase.

Schematic: A schematic is a collection of electronic symbols and virtual wires that represent a real-world circuit.

Schematic Capture: Schematic capture is the process of electronically drawing and connecting your design in an electronic media (your PC and software).

Silkscreen: Silkscreen is usually a white ink that is applied to the circuit board after the solder mask. This layer is typically used to outline components, add reference designators, add general text, instructions, etc.

Solder Mask: Solder mask is applied to the outer surfaces of a circuit board to prevent corrosion of the exposed copper. Areas of the board that will be tin-plated do not receive solder mask.

Solder Side: Solder side refers to the back side of the circuit board. This is the side where the leads of components come through.

Solderability: The solderability of a circuit board refers to the ease at which the board will accept solder. We tin plate our bare copper circuit boards to improve its solderability, as well as increase its reliability. Bare copper will oxidize after time, making it hard to solder to.

Surface Mount: Surface mount components do not have leads that pass through a circuit board. They are soldered directly onto one layer of a circuit board. They are usually very small.

Symbol Name: See *Component Name*.

Thermal Relief: A four-spoked padstack used to isolate a pad from a large area of copper. This technique allows an assembler to use a much smaller wattage soldering iron to connect to the pad.

Through-Hole: Through-hole components have leads that pass through a circuit board. These leads pass through drill holes that can connect one side of the circuit board to the other.

Tin-Plate: The process of depositing tin on bare copper. This is done to protect the bare copper from oxidation and enhance its solderability.

Toner: Toner is the black material that the letters on a photocopy are made of. It is a fine powder that melts under extreme temperatures and then hardens to form a waterproof, opaque solid.

Trace: A trace is a "printed wire" that connects the pads together on a circuit board.

Via: A via is a connection from a trace on one side of a printed circuit board to the other. In our case, a via is made with a barrel.

Wetting: Wetting refers to the process of melting solder onto a surface and then removing as much of it as possible. Wetting of pads is usually done to enhance the solderability of a device later on in the board assembly (or stuffing) process.

Index